Modeling Density-Driven Flow in Porous Media

Springer

Berlin
Heidelberg
New York
Barcelona
Budapest
Hong Kong
London
Milan
Paris
Singapore
Tokyo

Ekkehard O. Holzbecher

Modeling Density-Driven Flow in Porous Media

Principles, Numerics, Software

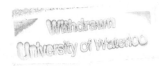

With 115 Figures and 52 Tables

 Springer

Dr. Ekkehard O. Holzbecher
Institute of Freshwater Ecology and Inland Fisheries (IGB)
Department of Eco-Hydrology
Rudower Chaussee 6A
Building 21.2

D-12484 Berlin
Germany

e-mail: Holzbecher@igb-berlin.de

ISBN 3-540-63677-3 Springer-Verlag Berlin Heidelberg New York

Library of Congress Cataloging-in-Publication Data

Holzbecher, Ekkehard, 1954-   Modeling density-driven flow in porous media: principles, numerics, software / Ekkehard Holzbecher.   p. cm.   Includes bibliographical references and index.
ISBN 3-540-63677-3 (hardcover)
1. Groundwater flow--Mathematical models. 2. Groundwater flow--Computer programs.
3. Porous materials--Mathematical models. 4. Porous materials--Computer programs. I. Title.
TC176.H647   1998   98-20099
551.49--dc21   CIP

Springer-Verlag Berlin Heidelberg 1998
Printed in Germany

The use of general descriptive names, registered names, trademarks, etc. in this publication does not imply, even in the absence of a specific statement, that such names are exempt from the relevant protective laws and regulations and therefore free general use.

Cover Design: design & production, Heidelberg
Typesetting: by author

SPIN: 10537790   30/3136 – 5 4 3 2 1 0 – Printed on acid free paper

Acknowledgments

This book would not have come into existence without the international HYDROCOIN project in which I had the pleasure being involved and it is a late result of discussions on the 'brine' test cases. The German team within HYDROCOIN was directed by Prof. Gerhard Memmert. Later I had the chance to extend my experience in a research project on seawater intrusion in the Nile delta aquifer led by Prof. Uwe Tröger within the 'Sonderforschungsbereich 69' (Geosciences in arid and semiarid climates). Further work on the subject especially on geothermal problems was enabled by the Ministry of Science, Technology and Education of Japan during a one year research stay at the Beppu Geophysical Research Institute of Kyoto University where I got the opportunity to cooperate closely with Prof. Yuki Yusa. Two years of funding by the 'Deutsche Forschungsgemeinschaft' (DFG) gave the final impulse for writing. At that stage support came from Prof. Gerd Holtorff and Prof. Wolfgang Kinzelbach, later from Prof. Asaf Pekdeger. Designing the final form of book and software I got help and advice from colleges at the Institute of Freshwater Ecology and Inland Fisheries in Berlin.

The author wishes to thank all those who provided help and input for this book in any way. A few I want to mention by their name: Dr. Gunnar Nützmann, Dr. Michael Kasch, Dr. Hans Jörg Diersch, Dr. Jörn Springer, Dr. Michael Thiele, Hans-Jörg Friedrich, Barbara Kobisch and Tom Cuson. My brother Harald Holzbecher I mention with special gratitude for updating the GeoShell graphical user interface even after having finished his thesis on GUIs in computer science. Special thanks to my family, my wife Susanne, daughter Gesa and son Gero, for understanding and encouragement.

The author wishes to thank in advance all those who respond with questions, critiques and tips.

Foreword

Density-driven flows in porous media appear in a wide variety of fields and are of growing interest. A prominent application is in groundwater quality protection against seawater intrusion and upconing of lower lying saline waters. With increasing water demand the control of salinity is one of the major tasks in sustainable groundwater management in many countries. Density contrasts caused by saltwater and the resulting flow patterns are also of relevance in the risk-assessment of salt-domes serving as host rock for repositories of hazardous wastes. In geothermal applications, on the other hand, temperature contrasts are the cause of density differences and the resulting thermal convection. Thermohaline flows which couple density effects of both origins may play a role in oil reservoir formation.

The book presented here addresses a wide scope of problems and provides the tools for tackling them. The theory is developed starting from an introductory level and proceeds up to advanced subjects concerning stability. The reader will certainly share my experience that density driven flow is rich in interesting phenomena such as fingering, bifurcation, and rotational flow. This abundance of structure is due to the non-linear nature of the governing equations. While for linear problems in groundwater flow and transport, numerical solutions and methods are well-developed and applied in a rather standard fashion today, non-linearity is the most thrilling subject for future research in modeling.

A numerical prototype code is supplied which allows to carry out a variety of numerical experiments in two vertical dimensions. This hands-on approach allows the reader to acquire familiarity with numerical solution methods, their convergence behavior and eventually their pathologies. The numerical solution of density driven flows in porous media is still far from being routine. Therefore practical experience with codes is highly welcome as a training for the critical eye. It is suggested to perform sensitivity studies both with respect to physical parameters and numerical parameters and discretization.

I wish the book success and its readers as much enjoyment as I had when reading and experimenting.

Zurich, in February 1998

Wolfgang Kinzelbach

Preface

'*Modeling density-driven flow in porous media*' as a multidisciplinary topic is of interest to a wide range of students, scientists and technicians.

Like 'density-driven flow' itself, this book is composed of an array of aspects. The book is intended not only for people at different levels, but for people from different disciplines: geologists, engineers, physicists, mathematicians. For any given reader, this means some aspects are neither relevant, interesting, nor understandable.

If you find yourself trapped in a scientific discourse that does not belong to your field and range of application, please don't worry! Geologists can skip all that stuff on bifurcations! Mathematicians have no need to go into the details of salt-dome formations!

The book can be broadly divided in four parts. Part One, chapters 1-4, provide the basics. Part Two, chapters 5-7, tackle convection. Part Three, chapters 8-10, deals with flow patterns induced by temperature differences (thermal problems). Part Four, chapters 11-14, examines salt concentrations as origin of flow patterns (saline problems).

Software is included on CD-ROM allowing the user to independently model problems of density-driven flow. The programs are much more than a demonstration of the textbook examples. There are some limitations, of course, and these are noted. A user-friendly interface helps the modeler with input data.

The practice of modeling density-driven flow still has obstacles, such as errors in input data, inconsistencies in the conceptual model, insufficiently small discrete parameters, insufficiently large hardware memory. Most of these obstacles are *not* specifically reported by the software, but they are noticeable when the output shows impossible patterns, the algorithm does not converge or the program crashes.

Coping with these situations is easier if the modeler has a well founded understanding of the processes and of the program structure. This book is intended to provide a convenient tool for gaining this understanding and for helping those who want to set up good models of density-driven flow with the FAST-C(2D) code or with other software.

Berlin, in January 1998, *E. Holzbecher*

Contents

INDEX

Symbols

The use of symbols for main variables is consistent throughout the entire text. They are listed in the following tables. In some subchapters symbols from the list are used locally in a different meaning.
The print mode for symbols is applied as follows:

- scalar variables are denoted in normal italic letters
- vector variables are denoted in bold small letters
- matrix variables are denoted in bold capital letters

Greek Symbols

α	thermal expansion coefficient [1/°K]
α_L	longitudinal dispersivity [m]
α_T	transversal dispersivity [m]
β	compressibility [1/Pa]
χ	automatic timestepping parameter [1]
Δt	timestep [s]
$\Delta\rho$	density change [kg/m^3]
ε	relative accuracy of a numerical method [1]
φ	porosity [1]
γ	ratio of heat capacities $((\rho c)_f/(\rho c)^*)$ [1]
ϑ	convection cell aspect ratio [1]
κ	time level weighting factor [1]
λ	thermal conductivity [W/m/°K]
μ	dynamic viscosity [Pa·s] [kg/m/s]
v	kinematic viscosity [m^2/s]
θ	normalized temperature or salinity [1]
ω	vorticity [1/s]
ρ	water density [kg/m^3]
$(\rho c)_f$	water specific heat capacity [J/m^3/°K]
$(\rho c)^*$	porous medium specific heat capacity [J/m^3/°K]
ψ	streamfunction (based on mass fluxes) [kg/m/s]
Ψ	streamfunction (based on volume fluxes) [m^2/s]
τ	overheat capacity [1]

Latin Symbols

c	concentration: volume fraction [kg/m^3]
c_m	concentration: mass fraction [kg/kg]
Cou_x, Cou_z	Courant numbers [1]
D	diffusivity [m^2/s]
$\mathbf{D}$	dispersion tensor [m^2/s]
e_z	unit vector in direction of gravity [1]
f_μ	correction factor due to viscosity change
f_ρ	correction factor due to nonlinear density change
g	acceleration due to gravity [m/s^2]
$\mathbf{g}$	vector in direction of gravity with length g [m/s^2]
H	height (vertical direction) [m]
I	hydraulic gradient [1]
j_c	diffusive flux of dissolved mass [kg/s]
j_h	diffusive heat flux [W/m^2]
$k, \mathbf{k}$	scalar, tensor of permeabilities [m^2]
k_x, k_z	components of diagonal permeability tensor [m^2]
L	length (horizontal direction) [m]
Neu	Neumann number [1]
Nu	Nusselt number [1]
p	pressure [Pa]
Pe_x, Pe_z	grid Peclet numbers [1]
q_c	salt sink / source [kg/s/m^3]
q_h	heat sink / source [W/m^3]
Ra	Rayleigh number [1]
R	retardation [1]
t	time [s]
T	temperature [°K]
$\mathbf{u}$	mean intersticial velocity [m/s]
$\mathbf{v}$	Darcy-velocity [m/s]
x	coordinate axis in horizontal direction [m]
z	coordinate axis in direction of gravity [m]

Units in SI-system; note that some variables are used in dimensionless form in the text as well.

1 Introduction

1.1 Density-Driven Flow

Fluid flow is classified as *density-driven* if the flow pattern is influenced by density differences in the fluid system. The necessary condition for this type of flow is that density differs from one location to another. This holds for the steady state, when the system does not change with time. If the flow is transient, additionally there are temporal changes of density.

If in an experimental set-up in a hydro-lab a fluid with constant density (let's say: water), is replaced by another with constant but different density (let's say: oil), this is in no way *density-driven*. There are density differences in the example, but these do not determine the principal features of the replacement. If both fluids mix, density-driven effects may emerge in the transition zone.

Density-driven flow may occur in all types of fluids. It can be observed in systems that contain water only. But there can be more than just the fluid. There may be fluid phases and non-fluid phases. Specifically, a solid phase may be present, which does not flow - more precisely: for which flow motions can be neglected, because they are too small or too slow.

This is the situation that exists in a *porous medium*. Flow in the pore space may be density driven and this flow is the subject of this book. It is outlined below that flow in porous medium obeys a certain law (see: Darcy's Law, chapter 3) This law needs to be seen in contrast to the motion of fluids that fill the entire space - those 'ordinary fluids', 'free fluids' with a 'pure fluid phase'.

The density that is referred to here, is *fluid density* ρ. Rock density may be relevant for certain processes in a porous medium and so could possibly influence flow, but this rare situation is not considered in this book. Fluid density is influenced mainly by temperature and salinity. There is another dependency pressure that is of minor importance.

It is not only the flow pattern that characterizes density driven flow. The distribution of density as a function of time and space is directly coupled with the flow. In systems with density-driven flow it is often not the density itself, which is studied, but the variable with the main influence on density. In a cooling or heating experiment, for example, the temperature determines the distribution of density within the system. If both fresh and saline water are involved, the distribution of salt concentration is important.

Is the occurrence of spatial - and temporal - changes of fluid density sufficient to classify a flow pattern density-driven? In fact this is not true: if different densities in a fluid system are observed, the flow is not necessarily density-driven: density differences may be too small to have a substantial effect on the movement of the fluid particles.

The opposite situation may occur: small differences in density may induce a flow pattern that is totally different. *Convection* is probably the best-known example: circulating convective flow can be observed in all types of fluids. A common situation can be experienced by everyone in a room, in which a heat-source is switched on: heated air will start to rise in the vicinity of the source and relatively cool air will sink down further away. A laboratory experiment in closed box can be set-up showing no flow. Just a little bit of heat or salt added may change the situation completely.

Some scientists use the term *variable density flow*. Although this is definitely correct, the term misses to highlight the interest on the changed flow pattern. *Buoyancy-induced flow* is another term closely related to the subject (Turner 1973, Gebhart e.a. 1988). In scientific literature both terms can often be exchanged in the context in which they appear. Nevertheless, some differences in the underlying ideas make me prefer '*density-driven flow*'.

Buoyancy is a force related to density differences. The density difference $\Delta\rho$ needs to be multiplied by g, the acceleration due to gravity, to get the buoyancy force. If there are no density differences ($\Delta\rho=0$), there is no buoyancy. If there are effects due to buoyancy, these can be called 'density-driven' as well.

Is it the more general concept to talk about *density driven flow*? Not in principle, because if there are density differences, then there is buoyancy as well. Nevertheless, with its origin in hydraulics, the term buoyancy stems from the idea of changing weight. It is strongly connected to the idea of motions in or against the direction of gravity. But far beyond that, the entire flow pattern in two or three space dimensions may change due to density differences. Spatial density changes may influence the flow pattern as a whole. There will be parts with upward flow; there will be parts with downward flow and there will be parts with almost horizontal velocity vectors.

Concerning the occurrence of density driven flow Gebhart e.a. (1988) write: '*such flows are found in the air circulation around bodies, in the enclosures we frequent, in cooking, in processing, in pools of water, and in the atmospheric, lake and oceanic circulation at every scale. They are found inside planetary bodies and are presumed to occur in and around celestial ones*'. Several industrial branches are concerned with these phenomena: metal casing, ceramic engineering, insulating materials, chemical engineering.

Just to list some more areas of application, let me as the author give some personal information. My interest in 'density driven flow in porous media' began in a project on the safety of a planned nuclear waste repository in a salt-formation. It was questioned whether and how high salt concentrations influence the flow in the overlying aquifer and thus affect potential flowpaths of nuclides to human environments. I was involved in a project on sea-water intrusion, which is a problem of increasing importance in many coastal regions of the world. I tackled geothermal flow in a Japanese hot-spring area and had to give an expertise on saltwater upconing. Nowadays I am still involved in cases, where temperature gradients play a role.

Some of these problems can be found in the book. Others are treated additionally, like the famous Bénard-type convection experiments in Hele-Shaw cells and boxes filled with sand or glass beats. Some others are not treated, but should be mentioned: the rise of a hot blob in a Hele-Shaw is treated experimentally and numerically by Elder (1967b). The SWIFT (1982) code for modeling density driven flow in porous media has been designed originally for the simulation of waste injection into deep saline aquifers, which has been treated later by Ward e.a. (1987) and Vincent e.a. (1992). Hossain/Weber (1985) study the influence of heat producing radioactive waste on flow in boreholes. Density-driven flow obviously may play an important role in the formation of oil sand deposits in sedimentary basins (Garven 1989). Molenaar (1993) models a situation where saltwater penetrates a fresh-water aquifer from the side. Oostrom e.a. (1992) observe the migration of a dense leachate plume from a repository through the groundwater. Increased values of groundwater temperatures due to various sources in highly-populated regions is of concern for E. Mages (1994). The study of Aszódi A. (1996) concerns the behavior of waste canisters in the case of a fire emergency.

In some application areas the literature is so vast that only few references can be given here. Hot or cold water can be stored in the subsurface pore space. This can be a convenient alternative to other technologies of energy storage; particularly when hot water is stored in summer and pumped in winter, or when cold water is stored in winter and pumped in summer. Examples of *seasonal energy storage* are treated by Tsang e.a. (1981) and Sykes e.a. (1983). Some references concerning natural flow in and exploitation of geothermal systems can be found in chapter 10.

Last not least it has to be noted that flow in submerged porous media is mainly density driven, because the hydraulic gradient at the surface (interface to the free fluid region) can usually be neglected.

The book treats density driven flow of miscible fluids. The non-mixing assumption is mentioned at some places only because it is sometimes used in research on miscible fluids - see: *sharp interface* approach in chapters 11 and 12. The assumption leads to a significant simplification of the analytical and the numerical description. Readers interested in density effects of really immiscible fluids are referred to the literature on multiphase flow and transport (Helmig 1997).

1.2 Modeling

What is a model? Generally speaking: '*a model is a representation of an object, system, or idea in some form other than the entity itself* ' (Leendertse 1981).

The scope of this book are computer models. Thus the form, in which a concerned model appears, is a program run on a computer. To be able to run the model, there are requirements on hard- and software. In order to perform a run on a computer, two things are needed: the code (program or software) and input-data. There are various procedures to enter input data, which are more or less interactive and this should not be discussed here.

There are various other *forms* in which a model may be *formulated*. There are (physical) models, in which reality is represented in a certain scale: flow models can be found in hydro-laboratories for example. The models, which are set up on computers are of totally different type. Aziz/Settari (1979) distinguish mathematical, numerical and computer models. They state: '*the physical system to be modeled must be expressed in terms of appropriate mathematical equations*'. More exactly: analysis as a branch of mathematics in fact offers the means to describe real systems.

The analytical formulation itself is based on a model on another level, which can be placed between reality and analysis (Fig. 1.1). Each of the natural sciences uses its own terms to represent a part of reality, whether it is physics, chemistry, biology, ecology, hydraulics, hydrology or whatever. Some termini may be used in different scientific fields - some are common to various disciplines. On this level on may call it a *conceptual model*.

Variables are used to describe a certain situation and distinguish its state from other possible ones. Basic laws define the interaction of the variables. The connection of the laws describing the potentially relevant processes usually leads to a set of differential equations. Generally the interest lies on changes in space and time and thus its partial differential equations that analytically describe a real system. For a complete description boundary and initial values are needed. For density driven flow in porous media the derivation of the differential equations can be found in chapter 3.

While differential equations connect *continuous* variables, a computer is only able to treat *discrete* variables. It is numerical mathematics which is concerned with *discretization*. Numerics is another branch of mathematics and therefore I avoid the term 'mathematical model'. Instead *numerical model* is an appropriate term. From various discretization methods, such as Finite Elements, Finite Volumes, Finite Differences, spectral methods, finally linear or nonlinear equations result. The solution of those equations belongs to numerics as well.

Another form of model is obtained when a numerical method is implemented on a computer. As mentioned above beyond numerics there are questions of hardware, software implementation, input data. There are various programming languages, which can be used to transform a numerical method in a computer algorithm. Input data can be entered on different levels of inter-activity. Graphical user interfaces (GUIs) can be used or not. There are lots of alternatives concerning preparation of input data (pre-processing) and the presentation of output data (post-processing). The diversity and complexity of this field is usually underestimated, but here is not the place to discuss these problems in detail.

One code is given on CD, called FAST-C(2D), which enables the user, to set up, run and explore his own models. The code is equipped with a GUI, which allows to input data easily. The reader may open example files, treated in chapters 5 to 14 and make his own changes or sensitivity analyses. For more details on FAST-C(2D) see chapter 1.4 and chapter 4.

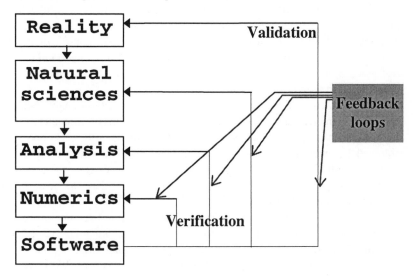

Fig. 1.1: Schematic view of model levels

Back to Fig. 1.1: below the level of reality, models can be found on all levels. In order to set-up a computer model, it is necessarily based on models on other levels, which in Fig. 1.1 are placed at an intermediate position. Physical, analytical and numerical forms have been mentioned - some others can be added. Geologists for example describe the subsurface in terms of rock or soil formations - that is one type of modeling. Hydrogeologists use a slightly different approach: they are mainly not interested in all rock types, but gather all formations with same hydraulic properties. Each scientific field and each level, which could be added to Fig. 1.1, has its own *termini technici*.

Abraham e.a. (1981) use a similar picture concerning modeling of flow and transport in surface waters. They introduce the concept of *'feedback loops'* between different levels. These loops are used to check, if certain steps from one level to the other are correct.

In that sense *code verification* is a feedback loop between the levels 'numerics' and 'software' on one side and 'analysis' on the other side. As a generalization verification may be understood as comparison of code output with generally accepted results; the latter could be results of other numerical models and codes.

Numerical codes are designed to solve problems which cannot be treated differently. If there would be an analytical solution, i.e. a direct solution at the level of mathematical analysis, there is no need to choose the more expensive way using a computer model. Verification is not done for the problem for which the software is to be used, but for a related problem, which can be solved analytically or for which well accepted results are available.

Often a solution can be obtained on the analytical level for a certain (mostly: idealized and simpler) model concept. Checking, if computer output for the simple model corresponds with the analytical solution, is a verification feedback loop for the numerical and the computer model.

Code verification has clearly to be distinguished from model *calibration* - the *'process of making the output of the model agree with a limited number of known responses of the problem'* (Abraham e.a. 1981). A feedback loop for a special model is referred to as calibration.

Most problems of practical concern can not be solved analytically: maybe because parameters are heterogeneously distributed, maybe boundary conditions are too complicated, maybe there are relevant nonlinearities. Feedback can be obtained by intercomparison with results from different codes designed for the same conceptual model. The basic idea is that solutions exist on the analysis level, but cannot expressed by analytical means. Nevertheless they can be approximated by numerical methods on a computer.

Modelers do not usually use different codes to tackle the same problem. They may compare results for similar exemplary problems, which are published in the scientific literature, with so called benchmarks. *Benchmarking*, as it has been done within international projects, such as HYDROCOIN (Cole e.a. 1988), is concerned with code verification. As models for benchmark cases are built on the same general analytical model - the feedback level is the same as for verification by analytical solutions. Benchmarking is an important part of *code testing*.

Validation is another often used term in modeling. *'A conceptual model and the computer code derived from it are validated when it is confirmed that the conceptual model and the derived computer provide a good representation of the actual processes occurring in the real system'* (IAEA 1988). Although there are various non-equivalent definitions of the term (Eisenberg e.a. 1988), most concepts have the representation of validation in common, which is given in Fig. 1.1: validation is the feedback loop from the computer back to the reality. Holzbecher (1997c) gives some further comments on validation.

1.3 Modeling Density-driven Flow in Porous Media

In a usual situation the modeler is allowed to treat flow and transport in separate steps. If there is any flow (i.e. non-zero velocity field) there is advection, which is one of the main transport processes (others are: diffusion, dispersion, decay and sorption). Thus transport is generally influenced by flow. Modelers take that into account, when they solve the flow problem first. When the velocity field is given, transport can then be simulated in a second step.

This procedure is allowed because transport has no influence on the flow. Often this is called the '*uncoupled situation*', which is incorrect, because there is a coupling. But the coupling is in one direction only - from flow to transport. This allows separate treatment in the solution procedure. If the interaction is in both directions, the method becomes inappropriate.

The distribution of a transport variable may affect flow, if one of the flow-relevant parameters is dependent on the variable. Important parameters are fluid properties and the most important of these is the density. If there are density-gradients in the system, flow will generally not be the same as in a constant-density situation. The model code needs to take the coupling into account, whenever flow and transport variables are calculated. Special codes need to be applied to manage the problem on the computer.

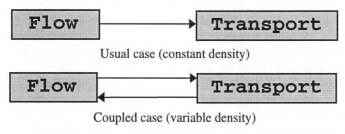

Usual case (constant density)

Coupled case (variable density)

Fig. 1.2: Schematic view of flow and transport interaction

The terms introduced in chapter 1.2, like verification and validation, are usually discussed for genuine flow or genuine transport in porous media. They are defined for the general process of modeling and thus are relevant for density driven flow as well.

Most problems in constant density flow or transport are connected with parameter identification, uncertainty and heterogeneity, i.e. problems in switching from reality to the conceptual model. The mathematical solution techniques for usual flow or transport - involving discretization and solution - are mainly straight forward (only the problem of numerical dispersion remains difficult).

The variable density field is complicated by the coupling between usually uncoupled processes. The analytical model is nonlinear in all cases and requires special numerical methods in discretization and solution. Analytical solutions do not exist. Thus the verification of the numerical codes in the above defined way is not possible. The feedback loop back to the level of mathematical analysis is hardly to perform. *'Performing a classical verification and validation exercise for coupled flow and transport models is an impossible task'* (Hassanizadeh 1990a). There remain two things that can be done: partial code verification and benchmarking .

Partial verification means that the code is verified in certain sub-problems, which can be tackled with code although it is not specially designed for them. I.e. benchmark and verification tests concerning usual flow or transport can be modeled. Density and other relevant properties, which may be influenced by the transport variable, can simply be set as constant. This should be possible in most variable density codes. In that way the transport and flow part of the program can be checked individually.

The coupling, for which those codes are specially designed, can be tested only by benchmarking. Few test-cases can be found in the scientific literature. The most relevant are reported in this book.

In the international HYDROCOIN project two benchmark cases have been defined and treated by several participants: the Elder experiment and the salt-dome problem. These are described here in chapters 9 and 13. Two benchmarks treating salt-water intrusion are described by Ségol (1993) *'to prove and improve models'*: Bear and Dagan's Hele-Shaw experiments and Henry's example (chapter 11). Others are recommended as benchmarks: steady convection in cold groundwater (chapter 6.3), saltwater-upconing below wells (chapter 12) and the geothermal flow problem (Yusa's example -chapter 13).

Special concern is given to the Bénard convection problem (chapters 5-7). For this problem there is a classical result concerning the margin given by the critical Rayleigh number. There are numerous other results: from the change of the critical margin due to box aspect ratios, viscosity dependencies, anisotropy to the existence of a second critical margin (the onset of oscillatory convection).

Beyond the examples studied in this book, the user may set up her/his own benchmarks for a certain application situation using the FAST-C(2D) code. The handling of the code for idealized situations is easy and 'user-friendly'.

Validation exercises are poorly presented in this book. Experimental data are rare. The Elder experiment (see chapter 9) is only partially good for validation, because the data presented in the papers seem to be numerically derived only. Recent experiments by a number of researchers are not taken into account here. The reader can get more information by Hassanizadeh (1990b), Moser (1995), Ackerer e.a. (1995).

Experiments reported in scientific literature can seldom be taken for a code validation, because the set-up and outcome is mostly not documented sufficiently. Validation is a multi-disciplinary task. Field and laboratory experiments need to be performed in close cooperation between modelers on one side and experimental scientists on the other side. If such a inter-disciplinary approach cannot be realized, the modeler must be less ambitious concerning validation. What modelers can do, is testing of hypotheses. Sensitivity analyses can be made to evaluate the relevance of parameters.

1.4 FAST-C(2D) Modeling Software

Most results of the classical benchmarks - Bénard convection, Elder experiment, Henry's problem, Yusa's problem, are based on the Oberbeck-Boussinesq assumption (see chapter 3) and use the streamfunction as flow variable. The code included on CD, FAST-C(2D), is constructed in the same way. The classical results can be checked on the numerical model directly.

There are difficulties in the verification or validation of a code, implemented on the general 3D-differential equations without any restricting assumptions. When a classical case is modeled, the difficulty is to explain occurring deviations from classical findings. The well accepted results of the last decades have all be obtained using simplifying assumptions.

Are the mentioned deviations an effect of the different treatment of validity/non-validity of the Oberbeck-Boussinesq assumption (chapter 3) or some other assumption? Are they an effect of the different representation of the flow field? Or are they an effect of a software bug, what to find out is the intention of a verification task? There is no easy answer at hand in the described situation.

In order avoid this, the FAST-C(2D) code has been developed intentionally, which is constructed on equations which are very near to those, for which the analytical results have been obtained. This enables the exploration of numerical effects.

There are some restrictions of models with FAST-C(2D). The code allows modeling in 2D vertical cross-sections only. The validity of the Oberbeck-Boussinesq assumption is generally taken for granted. Some other restrictions concern boundary conditions and grid generation.

On the other hand FAST-C(2D) is an ideal tool for beginners. A graphical user interface (GeoShell) provides a software environment that is easy to use. The input of data can be handled in a way, familiar the user from other WINDOWS software. Moreover the graphical representation on the computer display allows easy control and change of input data.

The FAST-C(2D) code is recommended for modelers, looking into density driven flow for the first time, as well as modelers, more deeply involved in that field - despite the above mentioned restrictions. The FAST-C(2D) results stands intermediate between analytical findings and outputs from elaborated models. It can be extremely useful in the process of verification of other codes.

2 Density and other Water Properties

Fluid properties which may need to be considered in studies of water flow include density, compressibility, viscosity, specific heat, heat capacity, thermal expansion coefficient, thermal conductivity and surface tension. Although the scope of this contribution is density-driven flow, other fluid characteristics need to be considered additionally. Because with density changes other properties change as well - temporally and spatially. Salt or heat variations, which cause density to change, induce variations on other parameters. The effects from density changes are enlarged or reduced by the change of the other properties. In some cases density-driven flow will be favored - in some cases disfavored, when the simultaneous variation of other properties is taken into account.

Generally, water properties change with temperature, salinity and pressure. Most papers concerning the subject provide the change of a parameter on one of these three variables only, making assumptions concerning the other two variables. Often the assumption is that both remain constant, but others can be found as well. In the following some dependencies are listed, as they can be found in the scientific literature.

2.1 Dependence on Temperature

Temperature typically changes in the environment and thus in the subsurface as well. Mostly two main zones can be identified there. In the upper part the temperature in the soil is basically determined by the temperature above the ground surface, i.e. by meteorological conditions. Short-time fluctuations (for example between day and night) can seldom be observed at the groundwater table. But seasonal changes can be clearly distinguished in near surface aquifers (see Fig. 2.1). Their amplitude gradually decreases with the distance from the ground surface.

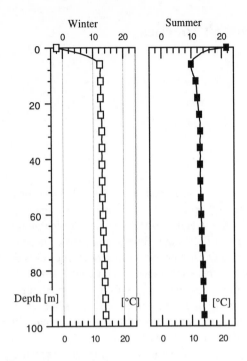

Fig. 2.1: Typical temperature profiles for winter and summer

In the lower part the influence from conditions above the ground are marginal and below a certain depth cannot be detected at all. If there are not any disturbances temperatures are almost steady state in this part.

The horizon between the two zones in near-surface aquifers in moderate climatic conditions can be found in a depth between 8 and 20 m. The illustration in Fig. 2.1 is based on measurements which have been made in January and July in northern Germany. The separating horizon between the two zones lies approximately at a depth of 13m below the ground surface.

In the lower zone, a gradual increase of the temperature with depth can be observed. The increase is determined by the heat flux from the interior of the earth. The so called *'geothermal gradient'* has been measured in many parts of the world and the mean value is approximately 3°C/100 m depth. Krige (1939) reports low gradients of less than 1°C/100m in several boreholes in South-Africa. In regions with geothermal activity the gradient may rise up to 6°C/100 m depth (Kitaoka/Kikkawa 1992). Deep subsurface temperatures increase to several hundred degrees centigrade. Under the high pressure conditions water is usually still saturated. Two-phase (liquid and gaseous) reservoirs or vapor-dominated conditions can only be found in regions with geothermal activity.

The geothermal gradient depends on the local geological environment in the subsurface. Convective motions of fluid in the pore space may change the temperature distribution substantially. Temperature profiles then depend basically on the local flow regime, which can be upward, downward or horizontal. The gradient may in some parts even show a change of sign, i.e. decreasing temperature with depth (see chapters 5 to 7).

2.1.1 Density

For fresh water, density approximations are given which are valid in different temperature intervals. The two formulae (2.1) hold for temperatures between 20°C and 250°C and between 100°C and 300°C. Both are valid for non-mineralized water at saturation pressure, i.e. under single phase conditions:

$$\rho = \begin{cases} 996.9(1 - 3.17 \cdot 10^{-4}(T - 298.15) - 2.56 \cdot 10^{-6}(T - 298.15)^2) \\ \qquad\qquad\qquad \text{for } 298.15 \prec T \prec 523.15 \\ 1758.4 + 10^{-3}T(-4.8434 \cdot 10^{-3} + T(1.0907 \cdot 10^{-5} - 9.8467 \cdot 10^{-9}T)) \\ \qquad\qquad\qquad \text{for } 373.15 \prec T \prec 593.15 \end{cases} \quad (2.1)$$

Input values for temperatures T need to be specified in °K. Both equations provide water density in the SI unit kg/m^3. The first one was taken from Wooding (1957), the second from Yusa/Oishi (1989). Fig. 2.2 gives a graphical representation of both approximations. There is a decrease of density of more than 30% over the whole temperature range considered here. Moreover it becomes obvious from the figure that, within the common range of temperatures, both curves coincide quite well.

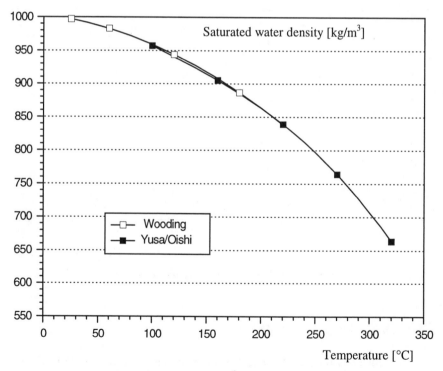

Fig. 2.2: Density of water [kg/m^3], dependency on temperature

For low temperatures, different approximations have to be used. The following two approximations can be found for water below 40° C:

$$\rho = 1000\cdot\left(1-\frac{(T-3.98)^2}{503570}\cdot\frac{T+283}{T+67.26}\right)$$

$$\rho = 1000\cdot\left(1-\frac{(T-3.98)^2}{508929.2}\cdot\frac{T+288.9414}{T+68.12963}\right)$$

(2. 2)

Density is calculated in SI-units and T has to be given in °C. The first formula has been proposed by Thiesen e.a. (1900), the second by Tilton e.a (1937). For temperatures below 20° C Bejan (1987) has used the simple quadratic polynomial function:

$$\rho = 1000\cdot(1-8\cdot10^{-6}(T-3.98)^2)$$

(2. 3)

Fig. 2.3 shows a comparison of the approximations. The well-known anomaly of water becomes clearly visible: fresh water has its highest density at 4°C.

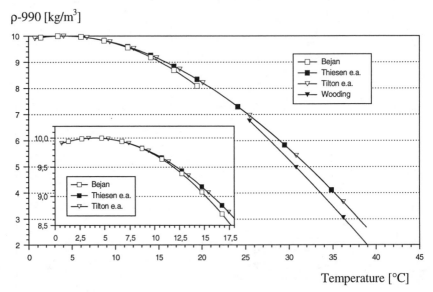

Fig. 2.3: Approximations for density of water below 40°C

2.1.2 Thermal Expansion Coefficient

Approximations for the coefficient of thermal expansion can be derived from the above formulas for density by using the following definition:

$$\alpha = -\frac{1}{\rho}\frac{\partial \rho}{\partial T} \qquad (2.\,4)$$

which holds under the assumption that changes in pressure and salinity are almost zero. From the polynomial functions for density, rational functions are obtained for the thermal expansion coefficient. For the two formulae (2.1), the resulting functions are plotted in the following figure. It can be noted that α increases by a factor of 10 within the whole temperature range.

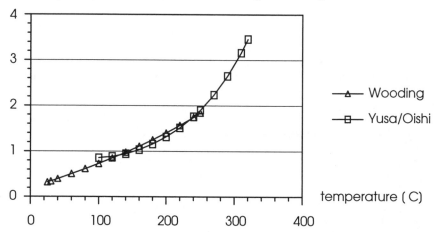

Fig. 2.4: Thermal expansion coefficient [10^{-3}/°C], dependency on temperature

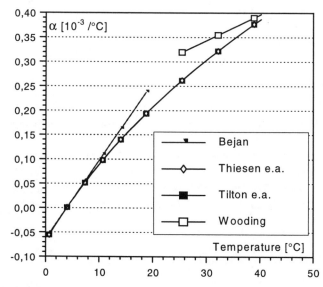

Fig. 2.5: Thermal expansion coefficient [10^{-3}/°C], dependency on temperature

The coincidence between both curves is less satisfactory than for the density approximations. Near 100°C, the values derived from the equation given by Yusa/Oishi (1989) are probably not reliable. Fig. 2.5 shows the functional dependence of α on temperature in the range below 40°C. The approximations have been derived from the different approximations given for density. For cold water below 4°C, α has negative values. Within this range it is a contraction coefficient, not an expansion coefficient.

2.1.3 Viscosity

For the temperature interval from 0°C to 100°C an invariant approximation is provided by Pawlowski (1991). Basic recalculation for T in unit [°K] yields the following formula:

$$\mu = 10^{-3}\{1 + 0.015512 \cdot (T - 293.15)\}^{-1.572} \tag{2.5}$$

The relative accuracy within the given interval is 0.2%.

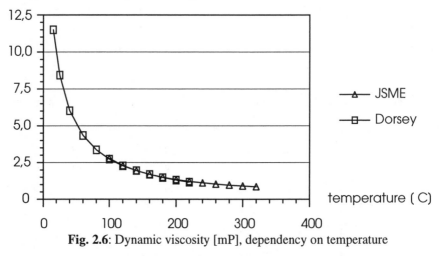

Fig. 2.6: Dynamic viscosity [mP], dependency on temperature

Approximations for dynamic viscosity of fresh water for higher temperatures can be found in the *'Steam Tables'* of the JSME (1968). The SI-unit for dynamic viscosity μ is [kg/m/s] or [Pa·s]. Values in SI-units need to be multiplied by 10 to obtain values in [P] (read: Poise). Viscosity in the unit [mP] is given by:

$$\mu = 241.4 \cdot 10^{-3} \cdot 10^{247.8/(T-140)} \qquad \text{for} \qquad 373.15 \prec T \prec 593.15 \tag{2.6}$$

Dorsey (1940) gives a formula for kinetic viscosity $v = \mu/\rho$ using the unit [cm^2/s]:

$$v = 0.332 / (T - 260.15) \qquad \text{for} \qquad 288.15 \prec T \prec 498.15 \tag{2.7}$$

For a comparison of both formulae, kinematic viscosity is multiplied by density taken from the formula (2.1a) given by Wooding (1957). Fig. 2.6 shows a good correspondence between both curves. Generally it can be observed that the change of μ is relatively dramatic. Viscosity decreases one order of magnitude within the 20°C-300°C temperature range.

2.1.4 Specific Heat Capacity

According to Yusa/Oishi (1989) specific heat capacity of fresh water at temperatures between 100° C and 320°C can be approximated by:

$$C = 3.3774 - 1.12665 \cdot 10^{-2} T + 1.34687 \cdot 10^{-5} T^2 \qquad (2.8)$$

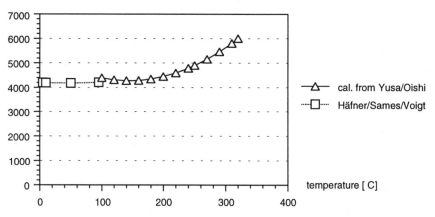

Fig. 2.7: Specific heat of water, dependency on temperature [J/kg/°K]

Unit is [cal/g/°K] and has to be multiplied by 4187.6 to obtain SI units [J/kg/°K]. Häfner e.a. (1992) give a few measured values in their publication which are shown additionally in Fig. 2.7.

Below 100°C, specific heat capacity seems to remain constant with a value of approximately 4200 J/kg/°K, while there is an increase of about 40% within the interval from 100°C and 300°C.

2.1.5 Thermal Conductivity

Thermal conductivity of water is also dependent on temperature. For temperatures between 0°C and 350°C the JSME (1968) according to Yusa/Oishi (1989) provides the formula:

$$\lambda_f = -922.47 + 2839.5\left(\frac{T}{T_0}\right) - 1800.7\left(\frac{T}{T_0}\right)^2 + 525.77\left(\frac{T}{T_0}\right)^3 - 73.44\left(\frac{T}{T_0}\right)^4 \quad (2.9)$$

T has to be given in [°K] and T_0=273.15°K. The physical unit for thermal conductivity is 10^3 W/m/°K. Some values for temperatures below 100°C can be found in Häfner e.a. (1992). A comparison (Fig. 2.8) shows that the latter data coincide quite well when the 100°C limit is approached. Obviously thermal conductivity of water has a maximum near 140°C. Maximum change within the given range is from 140°C to 359°C and amounts to 30%. For other temperature ranges, the change is much smaller.

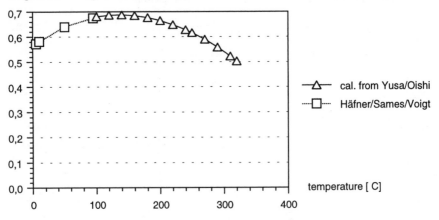

Fig. 2.8: Thermal conductivity of water [W/m/°C], dependency on temperature

In a porous medium, the thermal conductivity of the whole system containing fluid and solid material is relevant. The conductivity of rocks or soil is usually higher, ranging from 0.6 up to 7.7 W/m/°K for quartz (compare: Häfner e.a. 1992). If the temperature variation of rock conductivity is smaller than that of water, changes for the two-phase conductivity are smaller than those in λ_f.

2.1.6 Diffusivity

Thermal diffusivity of water can be combined from parameters as density, specific heat and thermal conductivity:

$$D_f = \frac{\lambda_f}{\rho_f C_f} \quad (2.10)$$

If the dependencies of the parameters are combined, the graph in Fig. 2.9 results. There is a small gap near 100°C, where the values have been calculated from data from different references. The main features can be observed clearly:

there is a maximum in thermal diffusivity of water near 160°C and maximum change is around 25%.

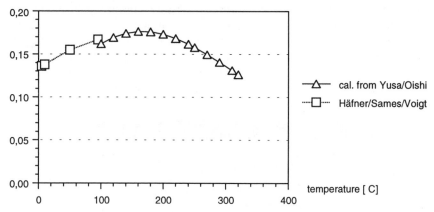

Fig. 2.9: Thermal diffusivity [cm²/s] of water, dependency on temperature

It should be noted that thermal diffusivity of combined porous medium and fluid can be calculated from the thermal conductivity of the two phase system:

$$D^* = \frac{\lambda^*}{(\rho C)^*}$$ (2. 11)

If changes of specific heat capacity and thermal conductivity of the solid material due to varying temperature are smaller than those of water, the variation of D^* will be smaller than that of D_f, too. Seipold (1995) measured thermal conductivity and thermal diffusivity of rocks in the earth crust. Pressure and temperature are chosen in a range valid for depths until 30km below the earth surface. Both diffusivity and conductivity decrease with increasing temperatures. For 1000°K, conductivity reaches the value of 2.0 W/m/°C, approximately. Diffusivity of rocks under the same condition is in the range of 0.3 mm²/s. Seipold (1995) summarizes: *'one has to keep in mind that the specific heat clearly increases with temperature. For this reason the temperature dependence is always much higher for the thermal diffusivity than for the conductivity'.*

2.2 Dependence on Salinity

Concentration in water is mostly given in physical units which relate mass of salt to volume of fluid, i.e. mg/l or g/m³. Mass fractions can be found where mass of salt is related to mass of fluid. The dimensionless unit is then often noted as ppt, ppm, ppb etc. Switching from one unit class to the other requires that fluid density be known. If density is equal to the fresh water reference value ρ_f=1000 kg/m³, 1 mg/l equals 1 ppm.

Salt concentration is mostly not measured directly. Electric conductivity per unit length is used as equivalent instead. Groundwater can be viewed as an electrolyte because nearly all its major and minor dissolved constituents are present in ionic form. The most common unit is [μS/cm][1]. By measuring electrical conductivity - or its reciprocal: electrical resistivity - an overall characteristic is determined that does not include information on the concentrations of different ions.

Category	TDS (mg/l)
Freshwater	$0\text{-}10^3$
Brackish water	$10^3\text{-}10^4$
Saltwater	$10^4\text{-}10^5$
Brine	$>10^5$

Table 2.1: Classification of water due to concentrations (Freeze/Cherry 1979)

Another such overall classification type is TDS (total dissolved solids). If resistivity is measured in μS/cm and is to be converted to TDS in mg/l, the conversion factor for natural waters lies usually in between 0.55 and 0.75 (Hem 1970). Grohman (1987) for saltwater determined the following relationships:

$$c[\mu S \, / \, cm] \equiv 5.48 + 1.43 \cdot c[mg \, / \, l]$$

$$\text{or} \qquad c[mg \, / \, l] \equiv -3.83 + 0.699 \cdot c[\mu S \, / \, cm] \qquad (2.\,12\,)$$

Stuyfzand (1989) recommends the following formula:

$$c_m[kg \, / \, kg] = 0.69778 \cdot 10^{-6} c[\mu S \, / \, cm] \qquad (2.\,13\,)$$

Table 2.1 reproduces water categories due to TDS, as proposed by Freeze/Cherry (1979). Tibbals (1990) splits the brackish water category into two parts: *slightly saline* for $1\text{-}3 \cdot 10^3$ mg/l and *moderately saline* for $3\text{-}10 \cdot 10^3$ mg/l. Water with salinity higher than $3.5 \cdot 10^4$ mg/l is classified as *briny*. Bortnik e.a. (1992) sets $2.4 \cdot 10^4$ mg/l as the margin between *brackish* water and *sea* water.

The classification of water concerning salinity always depends on the purpose for which the water is be used. As an example: sea water with a density of 1025 kg/m³ has a low salt concentration compared to saturated brine, which may have a density around 1350 kg/m³ (Reilly/Goodman 1985). But for common purposes even sea water is already too salty. It cannot be used for irrigation because most crops do not tolerate even slightly increased salinity levels. For drinking water supply and for most industrial purposes the salt content is too high. Its corrosive effects make it useless for cooling in thermal power plants.

---

[1] S=Siemens as SI unit for electrical conductivity is the reciprocal of Ohm, which is the unit for electrical resistance

2.2.1 Density

An approximation formula in terms of TDS mass fraction c_m and temperature in [°C] is mentioned by Stuyfzand (1989):

$$\rho = 1000 \cdot (1 + 0.805 c_m - 6.5 \cdot 10^{-6}(T - 4 + 220 \cdot c_m)^2) \qquad (2.14)$$

Density of surface sea water usually can be found in the range between 1022 and 1028 kg/m³ (Reilly/Goodman 1985). Mean sea water density is 1025 kg/m³. Characteristics of sea water are given in Table 2.2.

Mass fraction	Salinity c_m [ppt]	Chlorinity [ppt]	Density [kg/m³]
3.5	34.84	19.29	1024.5

Table 2.2: Measured characteristics for sea-water at 20°C (Reilly/Goodman 1995)

If a linear relationship between salinity and density is assumed, the coefficient of density variability is given as 700 kg/m³ (Ségol 1994); Cussler (1984) suggests 756 kg/m³.

Starting from the principle of mass conservation Herbert e.a. (1988) argue that in terms of mass fractions c_m the density of a mixture of fluids with densities ρ_0 and ρ_s is given by:

$$\rho^{-1} = (1 - c_m)\rho_0^{-1} + c_m \rho_s^{-1} \qquad (2.15)$$

Comparison with data measured at 20°C and 1 atm pressure shows that the relationship (2. 15) coincides, as well as the exponential fit:

$$\rho = \rho_0 \left(\frac{\rho_s}{\rho_0} \right)^{c_m} \qquad (2.16)$$

The comparison extends over the whole range of c_m, i.e. from zero concentration up to the solubility limit of 26%. A salt-related test case of the INTRAVAL project (INTRAVAL 1991 used) the following formula:

$$\rho = \rho_0 \exp(0.6923 c_m) \qquad (2.17)$$

where the concentration c_m denotes the mass fraction of NaCl, which is the only salt component here. The physical unit is determined by ρ_0, which is the fresh water density. Cordier/Goblet (1991) use $\rho_0 = 998.23$ kg/m³. It has to be noted that a linear fit is also a reasonable approximation. Herbert e.a. (1988) show that a linear relation holds for volume fraction c_v when the principle of adding volumes is assumed. Though the later is not perfectly true in nature, *'the volumes of salt solution and water are nearly additive when mixed'*.

2.2.2 Viscosity

The salinity dependence of dynamic viscosity can be given in terms of mass fraction as well. (Cordier/Goblet 1991) use the following equation:

$$\mu = \mu_0 (1.+1.85c_m - 4.1c_m^2 + 44.5c_m^3) \qquad (2.\,18)$$

μ_0=0.001002 Pa·s is viscosity of fresh water. Viscosity change is determined under isothermal conditions and certainly does not hold for the whole range of temperatures treated in chapter 2.1.

A more detailed approach is used by Olague e.a. (1991). The viscosity of a fluid containing N dissolved solids can be approximated as:

$$\mu = \mu_0 (1.+0.005\sqrt{\sum_{i=1,N} c_i} + \sum_{i=1,N} A_i c_i) \qquad (2.\,19)$$

where c_i denotes the concentration of ion i and A_i are temperature dependent coefficients.

If water contains less than 15 mol/l of dissolved ions, the square root in equation (2. 19) can be neglected. The main ions present are Na^+ and Cl^-. For these the coefficients A are $A(Na^+)$=0.08 and $A(Cl^-)$=0. Both are independent of temperature and the following simplified formula results:

$$\mu = \mu_0 (1.+0.08 \cdot c(Na^+)) \qquad (2.\,20)$$

2.3 Dependence on Pressure

2.3.1 Density

Dependence of density on pressure - according to INTRAVAL (1991) - is given by the formula:

$$\rho = \rho_0 \exp(4.5 \cdot 10^{-10}(p - p_0)) \qquad (2.\,21)$$

Physical unit for pressure is N/m^2. Liquids can usually be considered as incompressible, due to their low compressibility.

2.3.2 Compressibility

Compressibility at 1.02 atm according to Joseph (1976) is β=5.11·10^{-5} [1/bar]. In the publication compression is compared to contraction as result of temperature changes. Joseph (1976) takes the thermal expansion coefficient at 25°C, which is 2.64 10^{-4} [1/°C]. An easy calculation shows that the same change of density can be produced by a temperature difference of 1°C or a pressure difference of roughly five atmospheres. Thus changes in water density can be obtained much easier by varying temperature than by varying pressure.

In the following the main influences on water properties will be studied. As the influence of pressure on fluid properties is small relative to the influence of temperature and salt concentration, water is assumed to be incompressible. The assumption is not restrictive and is made in many studies on liquids.

3 Analytical Description

The complete analytical description for coupled flow and transport in a saturated porous medium is given through the set of differential equations derived in the following. Numerical calculations on computers are based on differential equations - or, in the terminology of chapter 1.2, a computer model depends on a model on the level of mathematical analysis.

There is a general distinction between parameters and variables. Parameters are assumed to be known. Parameters that will be dealt with are: permeability k, porosity φ, retardation R, diffusivity D, longitudinal and transversal dispersivity α_L and α_T, thermal conductivity λ, specific heat capacity ρc, and source- and sink-terms q. Some parameters already appeared in the previous chapter: the coefficients in the formulae are all parameters. If they are not known, when a model is to be set up for a specific site, they need to be determined experimentally or taken from the literature. This is a problem of the modeler, but not of the modeling code.

Variables, in contrast, are determined by a model. More precisely: the variables are calculated in the interior of a model area or volume, while at the boundaries more information needs to be given. Variables that can be found on the following pages are: density ρ, pressure p, salt concentration c, temperature T, streamfunction Ψ, velocity v and some others.

The aim of the rearrangements of equations is to reduce the number of main variables. Finally, a system of few (let's say M) differential equations remains for M variables. These are solved using relatively sophisticated numerical methods. The other variables can be determined easily in a straight forward manner.

The derivation is especially intended to lead to the set of differential equations used by the FAST-C(2D) code. The number of main variables is: $M=2$. There is one flow equation (for the unknown variable Ψ) and one transport equation (for the unknown variable θ). Density, viscosity and the velocity components can be calculated explicitly from Ψ and θ.

All variables and parameters can generally be understood as functions in time t and space - abbreviations of all main parameters are listed in the nomenclature. The distribution of the parameters in time and space is very different. Some, as permeability k or porosity φ, depend on the formation of the porous medium. They are independent of time and their values are spatially distributed in zones following the structure of the geological formation in the inhomogeneous case. They are constant in space in the homogeneous case. Others, like velocity $\mathbf{v}$, usually change from one place to the other and with time.

3.1 Basic Principles

In the absence of sources or sinks the principles concerning fluid flow can be expressed as:

$$\frac{\partial}{\partial t}(\varphi\rho) = -\nabla \cdot (\rho\mathbf{v})$$

$$\mathbf{v} = -\frac{\mathbf{k}}{\mu}(\nabla p - \rho\mathbf{g})$$

(3.1a,b)

As notation for the space derivatives in more than one dimensions, the *nabla* operator ∇ will be used in the book. This has the advantage that equations containing space derivatives can be written for systems in 1D, 2D or 3D equally well[1].

Equation (3.1a) is the analytical formulation of the conservation law for fluid mass. The second is an explicit formula for seepage velocity according to Darcy's Law[2]. The bold letter $\mathbf{k}$ denotes the permeability tensor. If $\mathbf{v}$ in (3.1a) is replaced by expression (3.1b), one equation results:

---

[1] The ∇-operator in two or three dimensions is defined by the following vectors:

$$\nabla = \begin{pmatrix} \frac{\partial}{\partial x} \\ \frac{\partial}{\partial z} \end{pmatrix} \quad \text{in 2D,} \qquad \nabla = \begin{pmatrix} \frac{\partial}{\partial x} \\ \frac{\partial}{\partial y} \\ \frac{\partial}{\partial z} \end{pmatrix} \quad \text{in 3D}$$

If ∇ is preceded by a dot ($\nabla \cdot$), the dot represents the vector multiplication. Thus ∇ has to be multiplied with the following vector. An example in 2D, when the vector is $v=(v_x, v_z)^T$:

$$\nabla \cdot \mathbf{v} = \frac{\partial}{\partial x}v_x + \frac{\partial}{\partial z}v_z$$

[2] Henri Darcy (1803-1858) in his publication '*Les fontaines publiques de la ville de Dijon, Libraire des corps impériaux des ponts et chaussées et des mines*', published in Paris in 1856 describes an experiment concerning water flow through a sand filled vertical column. He states the proportionality between the seepage velocity and the piezometric head gradient. The original experiment was performed in an idealized set-up with a homogeneous, isotropic porous medium and, saturated with a constant density, constant viscosity fluid. Several authors since then have generalized the law for inhomogeneous, anisotropic porous media partially saturated with variable property fluid. H. Darcy was an engineer, responsible for the water supply of the city of Dijon in France and did not derive any theoretical conclusions from his findings. Neither did he use the simple relationship, which he had

$$\frac{\partial}{\partial t}(\varphi\rho) = \nabla\cdot(\rho\frac{\mathbf{k}}{\mu}(\nabla p - \rho\mathbf{g}))\qquad(3.2)$$

Groundwater flow under usual conditions is modeled from this equation under the assumption that the parameters ρ, μ, φ and k do not change. Then only the pressure p remains as variable. Equation (3.2) is a differential equation for p and can be solved numerically if certain boundary and initial conditions are specified. (3.2) is the pressure formulation of the flow; other formulations can be used alternatively and are presented in following subchapters.

The permeability is usually not influenced by the flow and the above assumption is not a problem. Porosity may change under high pressure changes as an effect of compression or consolidation, but these processes are not of interest here. Additionally porosity may be influenced by dissolution or precipitation processes, but this also is rather an exception.

In contrast to the porous medium characteristics, changes in the fluid parameters are more important. Viscosity (μ) and density (ρ) may change due to salinity or temperature gradients. Then equations of state have to be taken into account in order to solve equation (3.2):

$$\rho(c, T, p)\qquad\qquad\mu(c, T, p)\qquad(3.3a,b)$$

Some of the formulations of state equations can be found in chapter 2. Two new variables come into the system by equations (3.3): now salt concentration c and temperature T need to be known in order to solve equation (3.2). The distribution of these variables is governed by the following principles. For the salinity field c the following equations hold:

$$\frac{\partial}{\partial t}(\varphi R\rho c) = -\nabla\cdot(\rho(\mathbf{v}c + \mathbf{j}_c)) + q_c$$
$$\mathbf{j}_c = -\varphi\mathbf{D}\nabla c\qquad(3.4a,b)$$

In the same way as equation (3.1a) for fluid mass, equation (3.4a) states that salt mass is conserved. On the left side of the equation the temporal change of mass is expressed. The first term on the right side, including the velocity $\mathbf{v}$, describes *advection* (not to be confused with convection - see chapters 5 to 7). The second term, including $\mathbf{j}_c$, stands for *diffusion* and *dispersion*. Source- and sink-terms for salt are taken into account by q_c. A *retardation factor R* is introduced in equation (3.4a), in order to take de- and adsorption processes into account. A detailed derivation of the retardation factor from the principle of mass conservation in a two-phase environment is given by Kinzelbach (1987) or by Holzbecher (1987b). In contrast to the transport of heavy metals or radionuclides R can mostly be omitted in problems of salt transport, because the main components of natural salt (Na^+, Cl^-) interact only slightly with the solid matrix and thus $R\approx1$.

---

found, for mathematical considerations - like Fourier did concerning heat conduction (see footnote 4).

Equation (3.4 b) is a generalized form of Fick's Law[3]. It describes mixing processes as an effect of concentration gradients. In Fick's formulation a scalar diffusivity D appears instead of the general *dispersion tensor* **D**. The general *dispersion tensor* is defined in a way to include the processes of molecular diffusion, longitudinal and transversal dispersion (see: Bear 1976, Scheidegger 1961). **D** is defined in detail by:

$$\mathbf{D} = \left(D_{ij} \right) = \left((D + \alpha_T u)\delta_{ij} + (\alpha_L - \alpha_T)\frac{u_i u_j}{u} \right) \tag{3.5}$$

where δ_{ij} is the Kronecker-symbol and indices $i,j \in \{1,2,3\}$ denote three coordinate axes.

Fig. 3.1 shows the various origins of dispersion, which are different from molecular diffusion. Most textbooks note the processes described in Fig.s 3.1a-c. The additional dispersive effect of temporal tidal fluctuations has been examined by Cooper (1959) and Dei (1978).

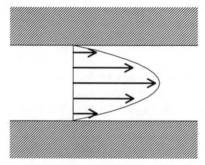

Fig. 3.1a: Origin of dispersion - changing velocity in pores

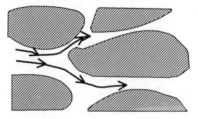

Fig. 3.1b: Origin of dispersion - different velocities in different pores

---

[3] Adolph Fick (1829-1901), professor for physiology in Zurich and Würzburg - the given equation is a reformulation of Fick's 1st Law, stating that the number of particles N crossing a surface A in the time interval t is proportional to the spatial gradient of the particle density $n=N/V$ (V for volume):

$$N = -t \cdot D \cdot A \cdot \partial n / \partial x$$

Fick's 2nd Law is the differential equation that follows from the conservation principle (Bergmann/Schäfer 1990) - a special case of the transport equation (3.7).

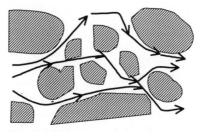

Fig. 3.1c: Origin of dispersion - changing direction of velocities in space

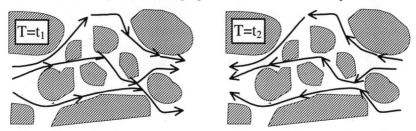

Fig. 3.1d: Origin of dispersion - changing direction of velocities in space and time

The coefficients of the dispersion tensor (3.5) are determined by changing absolute value and direction of velocities at the pore scale and at greater scales. For the regional scale the term *macrodispersion* can be found. It is determined at first rank by the inhomogeneities in the structure of the porous medium / media.

Prompted by theoretical considerations, Hassanizadeh/Gray (1979a, b, 1980) propose a further generalization:

$$\mathbf{j}_c(1+\beta|\mathbf{j}_c|) = -\rho \mathbf{D}\nabla c_m \qquad (3.6)$$

This implies in fact that the real mass flux is lower than the mass flux derived from the Bear-Scheidegger dispersion model. The stochastic approach by Welty/Gelhar (1991, 1992) results in a similar relationship, as shown by Moser (1995). One-dimensional experiments in a flow field in the direction opposite to gravity confirm the validity of a relationship as given by (3.6) (Moser 1995).

If velocity in the fluid flow equation is replaced with help of Darcy's Law, dispersive flux $\mathbf{j}_c$ in the first equation of (3.4) can be replaced by the second formula. It holds:

$$\frac{\partial}{\partial t}(\varphi R\rho c) = \nabla \cdot (\rho(-\mathbf{v}c + \varphi \mathbf{D}\nabla c)) + q_c \qquad (3.7)$$

This is the transport equation for a component moving in and with the fluid.

Another set of equations similar to equations (3.4) can be derived for the transport of temperature in a flow field. This set is:

$$\frac{\partial}{\partial t}(\varphi(\rho C)_f T_f) = -\nabla \cdot (\rho C \mathbf{v} T_f + \varphi \mathbf{j}_{hf}) + q_{hf} - q_{ex}$$

$$\mathbf{j}_{hf} = -\lambda_f \nabla T_f$$

(3.8a,b)

Note that the retardation parameter is omitted in the thermal case. The index f indicates that the parameter or variable is defined for the fluid phase only. The solid phase needs to be considered differently for heat flow, because there is additionally heat conduction through the porous medium. This flux has no analogue in solute transport.

q_{hf} represents heat sources or sinks. The additional sink/source-term q_{ex} denotes the exchange of heat between solid and fluid phases.

Equation (3.8a) is based on the energy conservation principle. The equation, which describes the entire energy balance in the fluid phase, is given by:

$$\frac{\partial}{\partial t}\left(\frac{\varphi\rho}{2}\mathbf{v}^2 + \varphi\rho e - \varphi\rho g z\right) = -\nabla \cdot \left(\varphi\rho\mathbf{v}\left(\frac{1}{2}\mathbf{v}^2 + h - gz\right) + \varphi \mathbf{j}_{hf}\right) + q_{hf} - q_{ex}$$

(3.9)

On the left side the temporal derivatives of kinetic, internal and potential energy are added. On the right side one finds the mechanical energy flux, heat flux $\mathbf{j}_{hf}$ and the heat exchange between fluid and solid phases. e stands for (specific) internal energy per unit mass and $h=e+p/\rho$ denotes (specific) enthalpy per unit mass. Following Kasch (1997), equation (3.9) can be simplified to (3.8a) using several thermodynamic relations. The derivation is based on the assumption that thermal compression or expansion can be neglected.

In order to close the system, additional equations for the heat flow through the porous medium itself need to be considered:

$$\frac{\partial}{\partial t}((1-\varphi)(\rho C)_s T_s) = -\nabla \cdot ((1-\varphi)\mathbf{j}_{hs}) + q_{hs} + q_{ex}$$

$$\mathbf{j}_{hs} = -\lambda_s \nabla T_s$$

(3.10a,b)

The index s indicates that the parameter or variable is defined for the solid phase only. The usual assumption is that both phases are in equilibrium. Physically speaking, the heat exchange between porous medium and fluid in the pore space is a fast process compared to the other transport processes. This assumption leads to a simplification of the system of differential equations.

Set $T=T_f=T_s$, $(\rho C)^* = \varphi \cdot (\rho C)_f + (1-\varphi) \cdot (\rho C)_s$ and $\mathbf{j}_h = \varphi \cdot \mathbf{j}_{hf} + (1-\varphi) \cdot \mathbf{j}_{hs}$ and add both equations (3.8a) and (3.10a) to obtain (3.11a).

$$\frac{\partial}{\partial t}((\rho C)^* T) = -\nabla \cdot (\rho C \mathbf{v} T + \mathbf{j}_h) + q_h$$

$$\mathbf{j}_h = -\lambda \nabla T$$

(3.11a,b)

Note that the heat exchange terms vanish because of their opposite sign in the original equations. λ is the thermal conductivity for the mixed system, representing the entire system with both solid and fluid phases. Density and specific heat capacity on the left side of equation (3.11a) are for the two-phase system. On the right side as coefficient in the advection term they appear as characteristics for the fluid phase only.

Terms for hot or cold sources are not included. The second equation (3.11b) is Fourier's Law[4] for porous media, i.e. flux and conductivity depend on both phases of the porous medium. If the diffusive flux term is replaced in the conservation equation one gets:

$$\frac{\partial}{\partial t}((\rho C)^* T) = \nabla \cdot (-\rho C v T + \lambda \nabla T) + q_h \qquad (3.12)$$

or (if temporal and spatial changes of $(\rho C)^*$ are neglected):

$$\frac{\partial T}{\partial t} = \nabla \cdot (-\gamma v T + D \nabla T) + \frac{q_h}{(\rho C)^*} \qquad \text{with} \qquad \gamma = \frac{\rho C}{(\rho C)^*} \qquad (3.13)$$

The whole set of equations (3.2), (3.3), (3.7) and (3.13), as given so far, describes variable density fluid flow coupled with salt and heat transport. The entire set is difficult to solve and most modelers make simplifications suitable for their specific problems.

There is one flow equation and there are two transport equations - one for salt concentration, the other for temperature. If the fluid density is influenced by gradients of both - temperature and salt concentration - certain special effects as fingering can be expected. In contrast to saline or thermal cases, where density is affected by only one of the two variables, this type of *thermohalin* problems is tackled by Nield (1968), Combarnous/Bories (1975), Tyvand (1980), Rubin/Roth (1983), Evans/Nunn (1989) and Sarker/Phillips (1992). More recent literature treats the problem under the term *double-diffusive convection*. Nevertheless, the porous medium case is seldom found in monographs (Brandt/Fernando 1995). The focus in this research lies on the interaction between flow, heat and salt patterns. Special effects, like salt fingers, have been discovered experimentally and in numerical models (Shen 1989, Murray/Chen 1989).

---

[4] Jean Baptiste Joseph Fourier (1768-1830) with his book '*Théorie analytique de la chaleur*' from 1822 sets the origin of thermodynamics - writes Truesdell (1971). Fourier not only stated the proportionality between heat flux and temperature gradient; moreover he included this relationship in a heat conservation law, leading to the differential equation for temperature

$$\rho C \partial T / \partial t = \nabla \cdot \lambda \nabla T$$

which is a special case of equation (3.8). In this heat conduction equation there is not to speak of a porous medium and there is not a flow field, which introduces advection as another process.

The following presents some simplifications of the entire set of equations. They are used by many authors in scientific publications. The most important assumption, the Oberbeck-Boussinesq assumption, concerns the relevance of density changes in different terms of the equations.

3.2 Oberbeck-Boussinesq Assumption

The Oberbeck-Boussinesq assumption leads to a substantial simplification of the differential equations. The most simple formulation is that changes in density can be neglected except from the buoyancy term $\rho\mathbf{g}$ in Darcy's Law (3.1b). Then equations (3.1) reduce to the system:

$$\frac{\partial\varphi}{\partial t} = \nabla\cdot\mathbf{v} \qquad \mathbf{v} = -\frac{\mathbf{k}}{\mu}(\nabla p - \rho\mathbf{g}) \qquad\qquad (3.14a,b)$$

Mostly porosity does not vary with time. Some exceptional cases, where this does not hold, are noted above. In the cases studied here the left side of (3.14a) vanishes and one can write as well:

$$\nabla\cdot\mathbf{v} = 0 \qquad\qquad (3.15)$$

If the explicit formula for velocity replaces $\mathbf{v}$ in the continuity equation, the pressure formulation for fluid flow is obtained:

$$\nabla\cdot\frac{\mathbf{k}}{\mu}\nabla p = \frac{\partial}{\partial z}(\frac{k_z}{\mu}\rho g) \qquad\qquad (3.16)$$

For isotropic case equation (3.16) has been used by Bear/Veruijt (1987). For the homogeneous and constant viscosity situation an even simpler formulation results that has first been studied by Knudsen (1962):

$$\nabla^2 p = g\frac{\partial\rho}{\partial z} \qquad\qquad (3.17)$$

The transport equations (3.7) or (3.13) simplify as well (source/sink terms are also omitted):

$$\frac{\partial}{\partial t}(\varphi Rc) = \nabla(\varphi\mathbf{D}\nabla c) - \nabla(\mathbf{v}c)$$

$$\frac{\partial T}{\partial t} = \nabla(D\nabla T) - \nabla(\gamma\mathbf{v}T) \qquad\qquad (3.18a,b)$$

The differential rule $\nabla(\theta\mathbf{v}) = \mathbf{v}\nabla\theta + \theta\nabla\mathbf{v}$ [5] lets the two advection terms be simplified by using equation (3.14a):

$$\frac{\partial}{\partial t}(\varphi Rc) = \nabla(\varphi\mathbf{D}\nabla c) - \mathbf{v}\nabla c$$

$$\frac{\partial T}{\partial t} = \nabla(D\nabla T) - \mathbf{v}\nabla(\gamma T)$$

(3.19a,b)

I want to note here that the arguments to derive the simplified set of equations are very different in the literature. Chandrasekhar (1961) argues that for most fluids, liquids or gases, the expansion factor is in the range of 10^{-3} to 10^{-4}. This range is very small and thus changes of ρ can be ignored where volume expansion or contraction is concerned.

Joseph (1976) provides an interesting overview on the historical perspective of the assumption which is mostly named only after Boussinesq. His arguments show that the mentioned simplification step would be better attributed to Oberbeck (1879) who:

- used this technique considerably earlier than Boussinesq (1903) and
- used a more rigorous approach.

Boussinesq derived the simplified set of equations from a list of assertions. Oberbeck in his paper already followed a modern line, trying to identify perturbation parameters, depending on a small entity ε, which produce the new set of differential equations in the limit $\varepsilon\rightarrow 0$. Joseph (1976) states that the Oberbeck-Boussinesq equations '*have not been completely justified*', although some work has been done by Mihaljan (1962) and Fife (1970). Joseph concludes, '*that there is no special reason besides our lack of proofs to doubt the validity of the nonlinear OB-equations*' (referring to the Oberbeck-Boussinesq-equations (2.12) and (2.15)).

It should be remarked that the subject is discussed with a view on the Bénard convection in pure fluids. The point, made in the argumentation, concerns the porous media as well:

1) the neglect of ρ-changes, where transport description and volume expansion are concerned, is identical for both cases even in its analytical form,
2) the buoyancy term, appearing in Darcy's Law, cannot be neglected as model results show; the resulting flow regimes may be totally different, if density changes in Darcy's Law are taken into account or not (see examples below).

---

[5] Note that the ∇-operator is used here for different analytical terms, depending if it is applied on a vector or a scalar variable. Thus it represents *divergence* in the first and last appearance of the formula, but *gradient* in the first term on the right hand side. This notation is used without any remarks everywhere else in this volume.

From the viewpoint of a finite volume numerical approach, an argument for the validity of the Oberbeck-Boussinesq assumption is that discrete values of densities should be interpreted as mean values for an entire volume. Within each volume or block the value of density used for the calculation of mass or heat flux is the same and thus can be canceled, for it is a multiplier in each equation of the finite formulation.

The system can be simplified further, using some often fulfilled assumptions. If the spatial and temporal variations of the specific heat capacity ratio and of the retardation can be neglected the equations (3.19) reduce to:

$$\varphi R \frac{\partial c}{\partial t} = \varphi \nabla (\mathbf{D} \nabla T) - \mathbf{v} \nabla c$$
$$\frac{1}{\gamma} \frac{\partial T}{\partial t} = \frac{1}{\gamma} \nabla (D \nabla T) - \mathbf{v} \nabla T$$

(3.20a,b)

3.3 Hydraulic Head Formulation

Most codes for groundwater models use hydraulic head h as the variable which describes flow. If z is measured positive in direction of gravity, hydraulic head is defined as

$$h = \frac{p}{\rho g} - z$$

(3.21)

For constant density fluids Darcy's Law defined in terms of h becomes simply

$$\mathbf{v} = -\mathbf{K} \cdot \nabla h \qquad \text{with} \quad \mathbf{K} = \frac{\rho g}{\mu} \mathbf{k}$$

(3.22a,b)

This is in fact the classical formulation of Darcy's Law (see footnote 2). Note that equation (3.22a) does not hold for the variable density case. Using the definitions (3.21) and (3.22b), the following can be derived from the generalized equation (3.1b):

$$\mathbf{v} = -\frac{\mathbf{k}}{\mu} (\nabla \rho g (h+z) - \rho g \mathbf{e_z}) = -\frac{\mathbf{k}}{\mu} (\rho g \nabla h + (h+z) g \nabla \rho) = -\mathbf{K} \nabla h - \frac{\mathbf{k}}{\mu} (h+z) g \nabla \rho$$

(3.23)

Another generalization of Darcy's Law for variable density can be formulated in terms of fresh water head h_f and fresh water conductivity $\mathbf{K_f}$ as:

$$h_f = \frac{p}{\rho_0 g} - z \qquad \text{and} \qquad \mathbf{K}_f = \frac{\rho_0 g}{\mu} \mathbf{k} \qquad (3.24\text{a,b})$$

Then holds:

$$\mathbf{v} = -\mathbf{K}_f (\nabla h_f - \frac{\rho - \rho_0}{\rho_0 g} \mathbf{g}) \qquad (3.25)$$

Using this formulation of Darcy's Law in equation (3.1) leads to a differential equation for the unknown variable h_f - for *'equivalent freshwater head'* (Senger/Fogg 1990) instead of pressure p. FEFLOW is one of the codes which are built that way (Kolditz 1994). The formulation with h_f has the advantage that fresh water hydraulic conductivity is needed as input parameter; most modelers are familiar with those. A disadvantage is that equivalent hydraulic head is a theoretically introduced variable, which cannot be measured directly. Measured values, such as pressures, used in boundary conditions or for calibration, need to be converted into the artificial variable h_f.

3.4 Streamfunction Formulation

In the following for 2D situations the streamfunction is used to describe flow. Under the condition $\nabla \cdot \mathbf{v} = 0$ the streamfunction Ψ in a (x,z)-coordinate system is given by the two defining equations:

$$\frac{\partial \Psi}{\partial x} = v_z \qquad\qquad \frac{\partial \Psi}{\partial z} = -v_x \qquad (3.26\text{a,b})$$

The unit of streamfunction in this formulation is *area/time* - in a system of unit width streamfunction differences denote volumetric fluxes between two streamlines. Formulae (3.26) are the 2D-formulation of the general 3D-equation $\mathbf{v} = \nabla \times \mathbf{\Psi}$. In 2D only one component $\Psi \equiv \Psi_y$ of the streamfunction vector $\mathbf{\Psi}$ is of interest: that is the one perpendicular to the considered plane (Peyret/Taylor 1985). The advantage in the 2D case is that one scalar variable[6] describes the flow field entirely. In 3D this advantage is not given: $\mathbf{\Psi}$ is a 3D-vector function.

Frind/Matanga (1985) show that (3.26) can be derived from the postulation that the vector of flux is tangent to the streamlines. Note that the use of both equations (3.26a,b) with a different sign introduces another streamfunction. In fact most publications take the sign conventions as given above.

---

[6] A scalar variable relates just *one* value to every point in the model region - in contrast to a vector function.

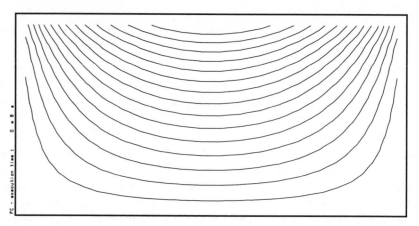

Fig. 3.2: Equally spaced streamlines for the Tóth (1962) case (chapters 10 and 13)

The physical unit of streamfunction is: length^2/time. In the 2D case Ψ is a discharge through a unit cross-section. If Ψ at two locations differs by $\Delta\Psi$, this is the numerical value for the discharge flowing between these locations. This property is often utilized in graphical representations when streamlines with equal spacing are chosen. The same amount of fluid flows within each *streamtube* (see Fig. 3.2). The velocities are high, where equally spaced streamlines are narrow. Velocities and flow rates are relatively low, where streamlines remain remote. Note that this nice feature is only applicable if lengths of the model region are shown in their real proportions.

In density driven problems Ψ is used to describe the steady state as well transient problems. Although the flow equation does not contain a time dependent term, the problem as a whole may be transient. Then the Ψ-plots visualize the flow pattern at certain times, i.e. the flow vectors are tangent to the streamfunction contours. But Ψ changes in time, the contours and flow vectors as well and the flowpaths of the particles with them.

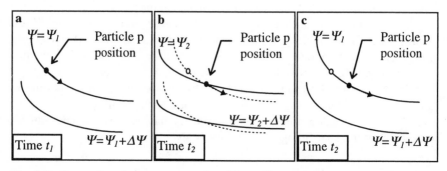

Fig. 3.3 : Streamfunction contours and pathlines (lines: current contours; dotted lines: old contours

Fig. 3.3 shows a the path of a particle, which has its position on the contour, where the streamfunction at time t_1 takes the value Ψ_1 (figure a). At a certain time $t_2 > t_1$ the position of the particle has changed. In the general transient case the flow pattern has changes as well, and the particle comes to lie on a contour for a different value Ψ_2 and a different shape and course (figure b). Thus at that point the pathline is headed into a different direction. The further course of the particle will be different, as if it would follow the original contour, on which is would stay in the case of steady state flow (figure c). Then pathlines and streamfunction contours are identical; these are *streamlines*.

An overview on the streamfunction concept is given by Zheng/Bennett (1995). Note that absolute values of streamfunction are arbitrary. For the modeler that means that she/he has to specify all values in relation to a reference value. It is the modeler's choice to determine the position and value of that reference. Values at other positions along the boundaries are determined using the discrete form of equations (3.26):

$$
\begin{aligned}
\Delta\Psi &= v_z \cdot \Delta x \qquad && \text{in x - direction} \\
\Delta\Psi &= -v_x \cdot \Delta z \qquad && \text{in z - direction}
\end{aligned}
\qquad (3.27)
$$

where $\Delta\Psi$ denote fluxes per unit width.

Frind/Matanga (1985) also discuss the anisotropic case without principal directions - for the constant density case only. The derivation is valid for the variable density case as well. For the 2D-case using equations (3.26) and Darcy's law, the following differential equation can be derived for the streamfunction:

$$
\frac{\partial}{\partial x}\left(\frac{\mu}{k_z}\frac{\partial\Psi}{\partial x}\right) + \frac{\partial}{\partial z}\left(\frac{\mu}{k_x}\frac{\partial\Psi}{\partial z}\right) = \frac{\partial}{\partial x}\left(\frac{\mu}{k_z}v_z\right) - \frac{\partial}{\partial z}\left(\frac{\mu}{k_x}v_x\right) = -g\frac{\partial\rho}{\partial x} \qquad (3.28)
$$

The mixed derivative terms of pressure vanish and a simple equation for the streamfunction remains, where the density on the right side stems from the buoyancy term. The equations (3.20) and (3.28) are a set of three differential equations describing flow of water with variable properties. It is completed by the explicit formulae in (3.26) and (3.3). This set was derived with no assumptions about the specific parameter dependencies of ρ, μ and D.

The streamfunction defined in terms of flux, as introduced by equations (3.26), necessarily fulfills the continuity equation $\nabla \cdot \mathbf{v} = 0$. Evans/Raffensperger (1992) recommend the use of a streamfunction ψ based on mass fluxes.

$$
\frac{\partial\psi}{\partial x} = \rho v_z \qquad\qquad \frac{\partial\psi}{\partial z} = -\rho v_x \qquad (3.29)
$$

In analogy to the treatment of the streamfunction Ψ based on volume flux, the following differential equation for ψ can be derived:

$$\frac{\partial}{\partial x}\left(\frac{\mu}{k_z \rho}\frac{\partial \psi}{\partial x}\right)+\frac{\partial}{\partial z}\left(\frac{\mu}{k_x \rho}\frac{\partial \psi}{\partial z}\right)=\frac{\partial}{\partial x}\left(\frac{\mu}{k_z}v_z\right)-\frac{\partial}{\partial z}\left(\frac{\mu}{k_x}v_x\right)=-g\frac{\partial \rho}{\partial x} \qquad (3.30)$$

Note that the use of the streamfunction ψ based on mass fluxes is not connected on the validity of the -Oberbeck-Boussinesq assumption. It holds:

$$0 = \nabla \cdot (\rho \mathbf{v}) = \rho \nabla \cdot \mathbf{v} + \mathbf{v}\nabla \rho \qquad (3.31)$$

and instead of $\nabla \cdot \mathbf{v} = 0$ follows.

$$\nabla \cdot \mathbf{v} = -\frac{\mathbf{v}\nabla \rho}{\rho} \qquad (3.32)$$

Senger/Fogg (1990) use freshwater hydraulic conductivity as defined in equation (3.24b) to derive:

$$\frac{\partial}{\partial x}\left(\frac{1}{K_z}\frac{\partial \Psi}{\partial x}+\frac{\rho-\rho_0}{\rho_0}\right)+\frac{\partial}{\partial z}\left(\frac{1}{K_x}\frac{\partial \Psi}{\partial z}\right)=0 \qquad (3.33)$$

This is equivalent to equation (3.28).

3.5 Vorticity Equation

The streamfunction formulation in free fluids is mostly used in combination with a vorticity equation. The vorticity vector $\boldsymbol{\omega}$ is defined as:

$$\boldsymbol{\omega} = \nabla \times \mathbf{v} \qquad (3.34)$$

where $\nabla \times$ is the so called *curl operator*, describing the *rotation* of a vector field. For 2D flow of free fluids the vorticity vector is always normal to the plane of flow. It can be shown, that a continuity equation holds for the scalar ω (Peyret/Taylor 1985). To derive that equation, the curl operator is applied to the Navier-Stokes equation. In the case of a porous medium the same operation is applied on Darcy's Law (3.1b):

$$\omega = (\nabla \times \mathbf{v})_y = \left(\nabla \times \left(-\frac{k}{\mu}(\nabla p - \rho \mathbf{g})\right)\right)_y =$$

$$= \left(-\frac{k}{\mu}\left(\overbrace{\nabla \times \nabla p}^{=0} - \nabla \times \rho \mathbf{g}\right)\right)_y = \frac{k}{\mu}(\nabla \rho \times \mathbf{g}) \qquad (3.35)$$

Note, that for some calculations it is required, that the system is isotropic, homogeneous and that there are no viscosity dependencies. A generalization for the case of variable viscosity is given by de Josselin de Jong (1960). Turner (1973) comments the result: '*Vorticity will be nonzero, whenever a variable density fluid is displaced from a state, in which $\nabla\rho$ is parallel to gravitation*'. As Combarnous/Bories (1975) already remark, the density gradient is parallel to the temperature gradient in thermal problems. If density change is caused only by salinity, $\nabla\rho$ is parallel to ∇c.

Turner adds: '*(in the simplest case) displacements of density surfaces away from horizontal produce vorticity*'. The latter becomes clearer in the following formulation:

$$\omega = \frac{kg}{\mu}\frac{\partial\rho}{\partial x} \qquad (3.36)$$

(3.36) is the *vorticity equation* in porous media. Garven (1989) refers to equation (3.36) as conservation of vorticity, but (3.36) can be derived easily from the relation between Ψ and ω

$$\nabla^2\Psi = -\omega \qquad (3.37)$$

Equation (3.37) holds for 2D and 3D, for free fluids and in flow through porous media.

The vorticity equation can seldom be found in the analysis of porous media flow. The reason is surely, that a continuity type equation for ω does not follow from the usual formulation of Darcy's Law.

Phillips (1991) shows how flow pattern characteristics can be obtained from the vorticity equation. Nonzero vorticity is an indicator for rotational motions of the flow. It can be shown that for an ideal circular rotation vorticity is proportional to azimutal speed u_r divided by the half-radius $r/2$ (Phillips 1991).

Vorticity transport is closely associated with the slope of isopygnals (constant density contours). If density gradient is nonzero, there is a nonzero vorticity as an indicator for rotational flow. Table 3.1 provides an overview on basic characterizations.

Density gradient	Vorticity direction	Rotation sense
positive	positive	clockwise
negative	negative	counter clockwise

Table 3.1: Characterization of rotational flow by vorticity and density gradient

Isopygnals are identical to isotherms if density is effected only by temperature. The characterization of rotational flow on temperature gradients has to take into account that density decreases with increasing temperature in the usual temperature range. For a negative temperature gradient there is positive vorticity and clockwise rotation. For a positive temperature gradient there is negative vorticity and counter clockwise rotation. Fig. 3.4 illustrates a situation where fluid in a porous medium is locally heated from below. The slope of temperature contours (dotted lines) changes above the hot plume. Also vorticity direction and eddy rotation change. Fluid motions similar to the shown situation were studied by Elder (1967b) in an experimental set-up. The Elder experiment is tackled in chapter 9.

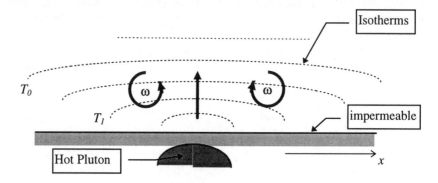

Fig. 3.4: Schematic view on relation between temperature gradients, vorticities and eddy rotation sense; for $T_0 < T_1$

Isopygnals are identical to isohalines if density is effected only by salinity. In saline cases the characterization of eddies can be taken from Table 3.1, when density gradient is replaced by salinity gradient. Fig. 3.5 illustrates the situation. The salt-dome problem with a similar concept is described and modeled in chapter 13.

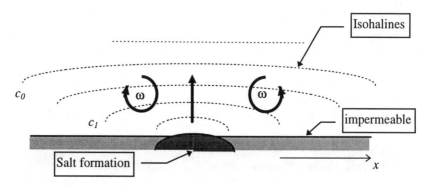

Fig. 3.5: Schematic view on relation between salinity gradients, vorticities and eddy rotation sense; for $c_0 < c_1$

In general density depends on both temperature and salinity (see chapter 2) and isopygnals are identical neither to isotherms nor isohalines. The characterization of flow patterns using vorticity can be extended to inhomogeneous and anisotropic porous media and to 3D situations (Phillips 1991). One significant result is that density-driven eddies in 2D horizontal layers do not occur.

3.6 Extended Oberbeck-Boussinesq Assumption

To derive a simpler equation which is valid as well in the case of variable viscosity, I introduce an argument in analogy to the Oberbeck-Boussinesq assumption: viscosity variations can be neglected. Then the whole equation can be divided by μ and as a result viscosity appears in the buoyancy term on the right side only:

$$\frac{\partial}{\partial x}\left(\frac{1}{k_z}\frac{\partial \Psi}{\partial x}\right) + \frac{\partial}{\partial z}\left(\frac{1}{k_x}\frac{\partial \Psi}{\partial z}\right) = -\frac{g}{\mu}\frac{\partial \rho}{\partial T}\frac{\partial T}{\partial x} \tag{3.38}$$

This is not the place to attempt a rigorous proof of that *extended Oberbeck-Boussinesq assumption*. The argument here is that local changes in viscosity do not play a significant role in equation (3.28). Viscosity may, in the extreme, change by one order of magnitude (see chapter 1.1.3), but that is true from one model region to the other. The local change of μ is much smaller and only in rare cases may exceed the (unknown) local inhomogeneity in permeability.

For the streamfunction based on mass fluxes holds with the same argument:

$$\frac{\partial}{\partial x}\left(\frac{1}{k_z\rho}\frac{\partial \psi}{\partial x}\right) + \frac{\partial}{\partial z}\left(\frac{1}{k_x\rho}\frac{\partial \psi}{\partial z}\right) = -\frac{g}{\mu}\frac{\partial \rho}{\partial T}\frac{\partial T}{\partial x} \tag{3.39}$$

The equations in this chapter are noted for density changes due to temperature gradients. For the saline case replace T by c.

3.7 Dimensionless Formulation

The dependency of fluid properties in most classical papers about density driven flow is studied for one variable only, either on T or on c. In order to come to a formulation which is valid in both cases, the transformations (3.40) have to be performed. The equal treatment for saline and thermal cases is only possible, if dispersion is assumed to be described by Fick's Law (3.4b) with a scalar coefficient D as well. The lengths are transformed using a typical length unit (here H):

$$x \rightarrow x / H$$

$$z \rightarrow z / H$$

$$t \rightarrow t \cdot D / H^2 \tag{3.40}$$

$$v \rightarrow \begin{cases} v \cdot H / D & \text{for the saline case} \\ v \cdot \gamma \cdot H / D & \text{for the thermal case} \end{cases}$$

$$\Psi \rightarrow \begin{cases} \Psi / D & \text{for the saline case} \\ \Psi \cdot \gamma / D & \text{for the thermal case} \end{cases}$$

The following notation for the transformed variables will be the same as for the original variables, as it is mostly clear which of both alternatives is meant. A new notation is introduced only for the transport variable: θ now stands for T or c - in the thermal or the saline case. Both transport equations (3.20), with the given transformations, come into the form:

$$\varphi R \frac{\partial \theta}{\partial t} = -\mathbf{v} \nabla \theta + \nabla^2 \theta \tag{3.41}$$

The coefficient function φR can be set to unity in the thermal case. A normalized temperature or concentration is used, in order to get dimensionless variables:

$$\theta = \begin{cases} (c - c_{\min}) / (c_{\max} - c_{\min}) & \text{for the saline case} \\ (T - T_{\min}) / (T_{\max} - T_{\min}) & \text{for the thermal case} \end{cases} \tag{3.42}$$

Using these transformations the flow equation (3.38) changes to:

$$\frac{\partial}{\partial x} \left(\frac{1}{k_z} \frac{\partial \Psi}{\partial x} \right) + \frac{\partial}{\partial z} \left(\frac{1}{k_x} \frac{\partial \Psi}{\partial z} \right) = \begin{cases} -\dfrac{gH}{D\mu} \dfrac{\partial \rho}{\partial \theta} \dfrac{\partial \theta}{\partial x} \\ -\dfrac{\gamma g H}{D\mu} \dfrac{\partial \rho}{\partial \theta} \dfrac{\partial \theta}{\partial x} \end{cases} \tag{3.43}$$

In homogeneous and isotropic porous media equation (3.43) can be multiplied by the permeability k. An additional assumption concerns the linearity of the state equation:

$$\rho = \rho_0 \pm \theta \Delta \rho \tag{3.44}$$

In the saline case, where density increases with salinity, the sign of the second term is positive. In the thermal case, with reference density ρ_0 for relatively cold water the density decreases with increasing temperature and the sign is negative. The differential equation for the streamfunction becomes:

$$\frac{\partial^2 \Psi}{\partial x^2} + \frac{\partial^2 \Psi}{\partial z^2} = \pm Ra \frac{\partial \theta}{\partial x}$$

$$\text{with } Ra = \begin{cases} \dfrac{k\Delta\rho g H}{D\mu} & \text{for saline cases} \\[3mm] \dfrac{k\gamma\Delta\rho g H}{D\mu} & \text{for thermal cases} \end{cases} \tag{3.45}$$

The dimensionless constant Ra is the so called *Rayleigh number*. Lord Rayleigh treated variable density flow of pure fluid systems in his classical publication from 1916 and derived the following dimensionless number as a characteristic of the system:

$$\bar{Ra} = \frac{g\Delta\rho H^3}{D\mu} \tag{3.46}$$

The porous medium Ra can be derived formally from the pure fluid $\bar{Ra}$ by multiplication with k/H^2. Lapwood (1948) was the first who adopted Rayleigh's analysis for the porous medium case (more details are given in chapter 5.2). For that reason the quotation of Ra in (3.45) as *'Lapwood number'* can be found in some publications. Some authors use the term *'filtration Rayleigh number'*. Although Lord Rayleigh, as mentioned above, did not study the porous medium and thus did not define a dimensionless constant for that case, the reference *'Rayleigh number'* has become common and is used in this publication as well.

The definitions of the Rayleigh number as given above are used in order to obtain the same flow equation for saline and thermal convection cases (see equation (3.45)). Lapwood for thermal convection used the definition

$$R\hat{a} = \frac{k\Delta\rho g H}{D\mu} \tag{3.47}$$

For the thermal case Elder (1967a, 1967b) uses a dimensionless vorticity equation (see chapter 9):

$$\omega = Ra \frac{\partial \theta}{\partial x} \tag{3.48}$$

in order to derive (3.45) with help of formula (3.37). De Josselin de Jong (1960) proceeds in the same manner.

If instead of Ψ the mass-based streamfunction, as defined by equations (3.29), is concerned, equation (3.45) has an analogue:

$$\frac{\partial}{\partial x}\left(\frac{1}{\rho}\frac{\partial \psi}{\partial x}\right) + \frac{\partial}{\partial z}\left(\frac{1}{\rho}\frac{\partial \psi}{\partial z}\right) = Ra \frac{\partial \theta}{\partial x} \tag{3.49}$$

In the cylindrical coordinate system the streamfunction can be defined by (Lamb 1963):

$$v_r = -\frac{1}{2\pi r}\frac{\partial \Psi}{\partial z} \qquad v_z = \frac{1}{2\pi r}\frac{\partial \Psi}{\partial r} \tag{3.50}$$

A streamfunction vector Ψ can be defined, which orthogonal to the flow plane has Ψ from (3.50) as only nonzero component. Note that Ψ does not fulfill the rotation relationship $\mathbf{v} = \nabla \times \Psi$ - in contrast to Ψ defined in chapter 3.4. In fact both streamfunctions can be transformed into each other using the multiplier $2\pi r$. The streamfunction, as defined in (3.50), in the literature is referred to as *Stokes' streamfunction* (Lamb 1963) or *Stokes's current function* (Sampson 1891). In this book I use the term streamfunction only. If the coordinate system is cartesian, Ψ is taken from (3.24). If the coordinate system is cylindrical, Ψ from (3.50) is meant.

In the radial cylindrical case the relation between streamline density and velocity, as noted above, does not hold as in cartesian coordinates. Velocities may be higher near to the inner radius, even if equidistant spaced streamlines remain remote. The characteristic remains true for the fluxes: the flux between equally spaced streamlines is the same.

Using the continuity equation (Lamb 1963, Diersch e.a. 1984)

$$\frac{1}{r}\frac{\partial}{\partial r}(rv_r) + \frac{\partial}{\partial z}v_z = 0 \tag{3.51}$$

and Darcy's Law

$$v_r = -\frac{k_r}{\mu}\frac{\partial p}{\partial r} \qquad v_z = -\frac{k_z}{\mu}\left(\frac{\partial p}{\partial z} - \rho g\right) \tag{3.52}$$

the flow equation results in the following formulation (Bear 1972):

$$\frac{\partial}{\partial z}\left(\frac{\mu}{k_r r}\frac{\partial \Psi}{\partial z}\right) + \frac{\partial}{\partial r}\left(\frac{\mu}{k_z r}\frac{\partial \Psi}{\partial r}\right) = -\frac{\partial}{\partial r}(\rho g) = -\Delta\rho g\frac{\partial \theta}{\partial r} \tag{3.53}$$

The transport equation becomes:

$$\frac{1}{r}\frac{\partial}{\partial r}\left(Dr\frac{\partial \theta}{\partial r}\right) + \frac{\partial}{\partial z}\left(D\frac{\partial \theta}{\partial z}\right) - v_r\frac{\partial \theta}{\partial r} - v_z\frac{\partial \theta}{\partial z} = \varphi\frac{\partial \theta}{\partial t} \tag{3.54}$$

or (for constant D):

$$\frac{\partial^2 \theta}{\partial r^2} + \frac{\partial^2 \theta}{\partial z^2} - \left(\frac{v_r}{D} - \frac{1}{r}\right)\frac{\partial \theta}{\partial r} - \frac{v_z}{D}\frac{\partial \theta}{\partial z} = \frac{\varphi}{D}\frac{\partial \theta}{\partial t} \tag{3.55}$$

3.8 Boundary Layer Formulation

A further simplification can be made for boundary layers. For vertical flow along a boundary the first derivative of the transport variable θ and the second derivative of Ψ in x-direction can be neglected. The flow equation (3.45) can be integrated once and with (3.41) the following system remains:

$$\frac{\partial \Psi}{\partial x} = \pm Ra \cdot \theta$$

$$R\frac{\partial \theta}{\partial t} = -\mathbf{v}\nabla\theta + \frac{\partial^2 \theta}{\partial x^2}$$

(3.56)

An equivalent set of equations is used for thermal flow by Ene/Poliševsky (1987). Besides the fact that $R=1$ for thermal systems, a different variable transformation is applied, which depends on the Rayleigh number itself.

Analogously, for horizontal boundary layers, neglecting derivatives in z-direction leads to the equations:

$$\frac{\partial^2 \Psi}{\partial z^2} = \pm Ra \cdot \frac{\partial \theta}{\partial x}$$

$$R\frac{\partial \theta}{\partial t} = -\mathbf{v}\nabla\theta + \frac{\partial^2 \theta}{\partial z^2}$$

(3.57)

Boundary layers will not be treated further in the following. Some solutions can be obtained analytically. The interested reader may look into Ene/Poliševsky (1987) for thermal flow in vertical layers and into Thiele (1993) for salt/freshwater mixing in horizontal layers.

3.9 Heat and Mass Transfer

The heat and mass transfer through a system is of profound interest for many application cases. It must be distinguished between conductive and convective systems, between local and total transfer.

Conductive heat flow is given by Fourier's Law (3.11b). For horizontal heat transfer through a no-flow system with length L, and for vertical heat transfer through a system of height H one obtains:

$$Q_{cond} = -\lambda \frac{T_{max} - T_{min}}{L} F \qquad \text{resp.} \qquad Q_{cond} = -\lambda \frac{T_{max} - T_{min}}{H} F \quad (3.58)$$

Both formulae quantify the total transfer through a surface with area measure F. The situation is more complex in a convective system, i.e. if there is flow along the boundary surfaces. The classical description is given by Nusselt[7] who wrote: *'Fließt an der Oberfläche eines festen Körpers eine Flüssigkeit oder ein Gas entlang, das eine andere Temperatur als die Körperoberfläche besitzt, so findet zwischen beiden ein Wärmeaustausch statt. Die an der Oberfläche erwärmten Flüssigkeitsteilchen werden durch die Strömung fortgeführt. Das über dem Körper in der Flüssigkeit entstehende Temperaturfeld hängt deshalb von dem Strömungsfeld ab und wird stark von ihm beeinflußt'[8]* (Nusselt 1944). For a given temperature field, according to Nusselt, the heat transfer through a closed boundary is given by:

$$Q_{conv} = -\lambda \int_F \frac{\partial T}{\partial n} df \qquad (3.59)$$

A generalization of the definition for varying thermal conductivity is given by:

$$Q_{conv} = -\int_F \lambda \frac{\partial T}{\partial n} df \qquad (3.60)$$

It is well known that heat transfer in a convective system is increased in comparison to the pure conduction case. The relative increase in recent literature is referenced as Nusselt number[9] and is thus given by:

$$Nu = \frac{Q_{conv}}{Q_{cond}} \geq 1 \qquad (3.61)$$

Nusselt numbers for general systems can be derived from the definitions just noted. For 2D-systems in cartesian coordinates with length L, height H and constant thermal conductivity the horizontal and vertical heat transfer, expressed in the Nusselt number, is given by:

$$Nu = \frac{L}{(T_{max} - T_{min})H} \int_0^H \frac{\partial T}{\partial x} dz \qquad \text{and} \qquad Nu = \frac{H}{(T_{max} - T_{min})L} \int_0^L \frac{\partial T}{\partial z} dx$$

$$(3.62)$$

Using the normalized space variables as defined in (3.40), and normalized temperatures (3.42) one obtains for the two special cases:

---

[7] Wilhelm Nusselt (1882-1957), Professor at Universities in Karlsruhe and Munich

[8] If a liquid or gas flows along a surface of a solid body, which has a different temperature than the solid surface, there is a heat exchange through heat conduction. The fluid particles, which are heated on the solid surface are taken away by the flow. The temperature field, which builds up above the body surface thus depends on fluid flow and is strongly influenced by the flow field.

[9] Nusselt himself introduced the *'Nusselt-Kennzahl'* in a different way (Nusselt 1944).

$$Nu = \frac{1}{H} \int_0^H \frac{\partial \theta}{\partial x} dz \qquad \text{and} \qquad Nu = \frac{1}{L} \int_0^L \frac{\partial \theta}{\partial z} dx \qquad\qquad (3.63a,b)$$

The analogue for the Nusselt number of the thermal case is the Sherwood number for the saline case.

4 Numerical Modeling (FAST-C(2D))

Modeling of density driven flow has to take into account as well flow as transport processes and their interaction. There are special techniques for some application cases, based on certain assumptions. Some of them will be discussed in following chapters in connection with the application field. The most advanced models are based on the formulation of flow and transport in differential equations, as they have been derived in the previous chapter.

There are no analytical solutions for non-trivial problems of density driven flow. For that reason solutions can be obtained by numerical methods only.

There are analytical solutions for the potential type equation describing flow in porous media - under certain circumstances as homogeneity and simple boundary conditions. Almost all books on groundwater flow provide some of them. O. Strack (1989) gives an excellent treatment and develops the method of *analytical elements* combining analytical solutions for problems of higher complexity. There are analytical solutions for the transport equation, describing diffusion, dispersion, advection and retardation - for certain conditions (see for example: van Genuchten 1981).

Solutions of the flow equation may be of interest in variable density flow as asymptotic cases, when density becomes constant and gradients become zero in the system. It is shown in chapter 13 that the Tóth-solution, which has been shown in Fig. 3.2., is a solution of the variable density salt-dome problem, when the entire system is filled with saltwater; i.e. when there are no density gradients any more.

If there are no analytical solutions, problems can mostly be treated by numerical calculations. Modeling or simulation on computers has become a common technique for solving flow or transport problems in porous media for constant density. The methods used to tackle the simpler situation can be taken over to the case of variable density flow.

The step from the differential equation to a formulation, which can be treated by computational methods, is noted as *discretization*. There are various common techniques and most of those, which are successfully used in other application fields have been adopted for modeling variable density flow. Thus finite differences (FD), finite elements (FE), finite volumes (FV) or the spectral method are used to discretize flow and transport equations.

The method of characteristics (MOC) for the treatment of solute transport has - to the knowledge of the author - been applied only by one group

(Sanford/Konikow 1985, Konikow e.a. 1997). The representation of solute concentration by the number of particles in cells, in fact includes significant uncertainties in the calculation of density. For density changes are the crucial for the changed flow pattern, these uncertainties limits the applicability of the method for the solution of the transport equation. The same argument holds for the random walk method (Kinzelbach 1987, Zheng/Bennett 1995). The method has been applied for pure advection-dispersion systems successfully, but cannot be found in literature on density-driven flow.

It is not intended to describe the details of the above mentioned discretization methods. As they are popular for the set-up of models in porous media systems, they are described in basic textbooks in that scientific field (Kinzelbach 1986, 1987, Bear/Verruijt 1987, Zheng/Bennett 1995, Holzbecher 1996). Manuals for certain codes describe additional information on the techniques (SWIFT 1984, SUTRA 1984, FEFLOW 1996).

In this chapter the focus lies on the presentation of the discretization methods, which are implemented in the FAST-C(2D) code, which is provided on the CD-ROM. Only the topic of the nonlinear coupling is treated more generally, because of its special relevance for density driven flow.

The FAST-C(2D)-code has been designed, implemented and improved permanently since early beginnings in 1986. The code can be applied to model density driven flow in porous media in 2D vertical cross-sections using the streamfunction formulation. FAST-C(2D) is based on a finite difference discretization of equations (3.18), (3.24) and (3.32).

There is an option to choose between steady-state and transient modeling. Another option concerns the physical dimensions: the dimensionless version allows the direct use of the Rayleigh number as input parameter directly.

Parameters as permeability, porosity, retardation, diffusivity, sink/source-rates, longitudinal and transversal dispersivity can be distributed within the model-region. The aquifer can be anisotropic. Grids are rectangular and can be irregular. Boundary conditions can be chosen as Dirichlet- or Neumann-type.

The FAST-C(2D) input shell, which provides an easy way to set up models for various situations. Especially useful is the input option for distributed parameters. They can be introduced into the model by mouse click. The graphical representation is an ideal tool for input, control and change of inhomogeneous parameters.

FAST-C(2D), in a modeling run on the computer, solves systems of two differential equations. Modeling is restricted to density variations caused by either temperature or salinity gradients. It is not (yet) designed to tackle thermohaline problems. FAST-C(2D) works with the scalar streamfunction; so its applications are restricted to 2D-vertical cross-sections. The code does not handle velocity and direction dependent dispersivity.

The code is based on the discretization of eq.s (3.26), (3.41) and (3.43), when the dimensionless formulation used. The same set of differential equations has been used to describe convective motions by Wooding (1957), Elder (1967a),

Elder (1967b) and Holzbecher (1991a). The same system has been used for numerical modeling of density driven flow by Henry (1960), Elder (1967a), Elder (1967b), Yusa (1983), Holzbecher (1991b), Baumann/Moser (1992), Holzbecher/Baumann (1994), Holzbecher (1995), Holzbecher/Heinl (1995), Baumann (1995) and Holzbecher/Yusa (1995).

The program allows the use of several functional relationships for density and viscosity dependence on temperature and salt concentration. Viscosity and density change may be linear by default. Viscosity may be constant or change exponentially with temperature. Density change is nonlinear for low temperatures. If constant density is selected, the flow problem and the transport problem become uncoupled (see chapter 1).

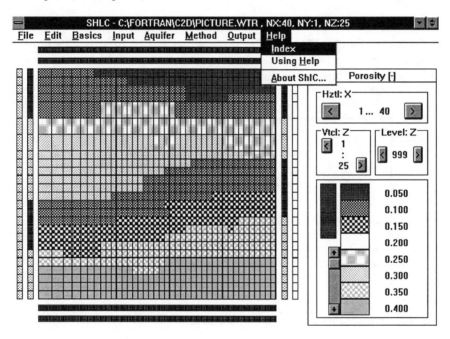

Fig. 4.1: Input Shell for FAST-C(2D) (Version 0.1)

Fig. 4.1 shows the graphical user interface[1] which the modeler uses for input, control and change of input data. A graphical representation of the model area provides an easy to handle tool for working with distributed parameters. Moreover several options concerning discretization alternatives, solution techniques and output can be set as well. The program is equipped with an online help[2] that provides all necessary information for working with the software.

---

[1] The program can be found on CD-ROM under the name *C2d_Shl.EXE*
[2] The help-file can be found on CD-ROM under the name *C2d_Shl.HLP*

Furthermore the modeling program FAST-C(2D) is called by selecting *'run'* in the pull-down menu. Then all input-data are stored on a file with the standard name *FLOAT.WTX*, from which FAST-C(2D) reads them again. An alternative way to set-up a model is, to write all input-data in an ASCII-file. An arbitrary (DOS or WINDOWS) editor can be used for that purpose. The input-data have to be structured following the input-file description in appendix I. Finally, the model run is started by executing FAST-C(2D) with the file-name as parameter.

4.1 Spatial Discretization

Grids used by the software are rectangular and may be irregular. There is no need for a grid preprocessor. The grid is completely characterized by the number of blocks in each of the two coordinate directions and two block-length vectors.

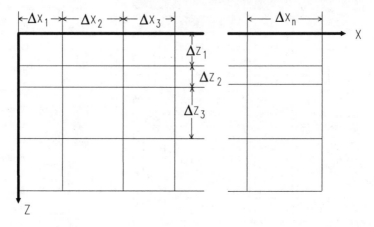

Fig. 4.2: Example for a grid used by FAST codes

First and second order derivatives are found in the differential equations. These are discretized following the usual finite difference approach. The second order terms in the transport equation are approximated by the standard scheme. In *x*-direction is:

$$\frac{\partial^2\theta}{\partial x^2} \approx \frac{\theta(x+\Delta x)-2\theta(x)+\theta(x-\Delta x)}{\Delta x^2}$$ (4.1)

For second order terms in the flow equation the following formula is applied:

$$\frac{\partial}{\partial x}\frac{1}{k_x}\frac{\partial \Psi}{\partial x} \approx \left(\frac{1}{k_x}\right)(x+\frac{\Delta x}{2})\frac{\Psi(x+\Delta x)-\Psi(x)}{(\Delta x)^2} - \left(\frac{1}{k_x}\right)(x-\frac{\Delta x}{2})\frac{\Psi(x)-\Psi(x-\Delta x)}{(\Delta x)^2}$$

(4.2)

Approximations (4.1) and (4.2) hold for equidistant grids; the approximations which are implemented in the code use the straight forward generalization for irregular grids. The coefficients in equation (4.2) have to be evaluated at the block-edges. For equidistant block-lengths the arithmetic mean of the permeability is used. If grid block lengths are different, the weighted mean is applied as follows:

$$\left(\frac{1}{k_x}\right)(x) = \frac{1}{\Delta x^- + \Delta x^+}\left(\frac{\Delta x^-}{k_x(x-\Delta x^-/2)} + \frac{\Delta x^+}{k_x(x+\Delta x^+/2)}\right)$$

(4.3)

The second order terms of the transport equation are treated in a similar way. The mean values of diffusivity have to be evaluated in analogy to equation (4.3) (replace the reciprocals of permeability by diffusivity!).

The 1st order terms in the transport equation have three choices: BIS (backward in space) or upwind scheme, BIS with truncation error correction, and CIS (central in space). For an equidistant grid BIS and CIS can be noted as:

BIS:

$$v_x\frac{\partial \theta}{\partial x} \approx v_x(x)\frac{\theta(x)-\theta(x-\Delta x)}{\Delta x} \qquad \text{if } v_x(x)>0$$

$$v_x\frac{\partial \theta}{\partial x} \approx v_x(x)\frac{\theta(x+\Delta x)-\theta(x)}{\Delta x} \qquad \text{if } v_x(x)\le 0$$

(4.4)

CIS:

$$v_x\frac{\partial \theta}{\partial x} \approx v_x(x)\frac{\theta(x+\Delta x)-\theta(x)}{\Delta x} - v_x(x)\frac{\theta(x)-\theta(x-\Delta x)}{\Delta x}$$

(4.5)

The advantage of the BIS-scheme is that it is more stable than the CIS-discretization. The disadvantage is an increased numerical dispersion. The first order term of the numerical dispersion can be derived by using the Taylor-series in equations (4.4):

$$v_x\frac{\theta(x)-\theta(x-\Delta x)}{\Delta x} = v_x\frac{\partial \theta}{\partial x} - v_x\frac{\Delta x}{2}\frac{\partial^2 \theta}{\partial x^2} + \Delta x^2(...)$$

(4.6)

The leading term of the truncation error is equal to a diffusion/dispersion -term (Lantz 1971). The amount of numerical dispersion is:

$$D_{num} = v_x\frac{\Delta x}{2}$$

(4.7)

Reducing the physical diffusion by this amount in each block results in the *truncation-error corrected* BIS method. Axelsson/Gustaffsson (1979) refer to this discretization as *modified upwind scheme*.

The coefficients in equations (4.4) and (4.5) are velocity components in block centers. These are determined as means from the parallel velocity components at block edges by using the following procedure:

$$v_x(x,z) = \frac{1}{2}\left(v_x(x - \frac{\Delta x}{2}, z) + v_x(x + \frac{\Delta x}{2}, z)\right)$$

(4.8)

The parallel velocity components at the edges which appear on the right side of equation (4.8) are obtained by the simplest finite difference formula which approximates (3.26b):

$$v_x(x,z) = -\frac{\Psi(x, z + \Delta z^+ / 2) - \Psi(x, z - \Delta z^- / 2)}{\Delta z}$$

(4.9)

A similar formula can be written for the z-components v_z as approximation of equation (3.26a).

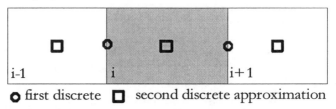

first discrete   second discrete approximation

Fig. 4.3: Scheme for calculating the vertical velocity component at the center of block i

Mixed second order terms in the transport equation are discretized using the formula

$$\frac{\partial^2 \theta}{\partial x \partial y} \approx \frac{1}{4\Delta x \Delta y}(\theta^{i+1,j+1} - \theta^{i+1,j-1} - \theta^{i-1,j+1} + \theta^{i-1,j-1})$$

(4.10)

This is a special case of the finite volume (FV) discretization

$$\frac{\partial}{\partial x}\left(D_{xy}\frac{\partial \theta}{\partial y}\right) = \frac{1}{\Delta x}\left\{ D_{xy}^{i+,j}\frac{1}{2}\left[\frac{\theta^{i+1,j+1} - \theta^{i+1,j-1}}{2\Delta y} + \frac{\theta^{i,j+1} - \theta^{i,j-1}}{2\Delta y}\right] - D_{xy}^{i-,j}\frac{1}{2}\left[\frac{\theta^{i-1,j+1} - \theta^{i-1,j-1}}{2\Delta y} + \frac{\theta^{i,j+1} - \theta^{i,j-1}}{2\Delta y}\right]\right\}$$

(4.11)

$$\frac{\partial}{\partial y}\left(D_{yx}\frac{\partial \theta}{\partial x}\right) = \frac{1}{\Delta y}\left\{D_{yx}^{i,j+}\frac{1}{2}\left[\frac{\theta^{i+1,j}-\theta^{i-1,j}}{2\Delta x}+\frac{\theta^{i+1,j+1}-\theta^{i-1,j+1}}{2\Delta x}\right]-\right.$$
$$\left. -D_{yx}^{i,j-}\frac{1}{2}\left[\frac{\theta^{i+1,j-1}-\theta^{i-1,j-1}}{2\Delta x}+\frac{\theta^{i+1,j}-\theta^{i-1,j}}{2\Delta x}\right]\right\}$$

(4.12)

when coefficients are evaluated at the block centers. In FAST-C(2D) the coefficients D_{xy} and D_{yx} are combined from dispersivities, diffusivity and velocity components. Velocity components are calculated according to formula (4.9).

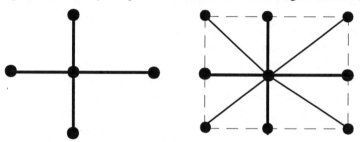

Fig. 4.4: Discretization stencils with 5 and 9 points respectively

Discretization errors have to be considered in irregular grids, because the approximations are not as accurate as in the equidistant case. Grid orientation may cause additional errors (de Vahl Davis/Mallinson 1972).

The main criterion for the stability and accuracy of the finite difference method applied on the transport equation is formulated using the grid-Péclet numbers[3]:

$$Pe_x = \frac{v_x \Delta x}{D} \qquad\qquad Pe_z = \frac{v_z \Delta z}{D} \qquad (4.13)$$

The criteria are:

$$Pe_x \le 2 \qquad\qquad Pe_z \le 2 \qquad (4.14)$$

It has to be noted that the grid-Péclet numbers change spatially and temporally (in transient models). Density-driven flow always shows heterogeneous velocity fields - the main origin for the change of Pe_x and Pe_z. But grid-Péclet numbers are altered, where diffusivity or grid spacing change.

As a combination of physical and numerical parameters, grid-Péclet numbers characterize a numerical model. In fact (4.14) are criteria to check if a grid is fine

---

[3] The name of the dimensionless number stems from the Péclet number, which is defined as the same combination of parameters, when the numerical parameters Δx and Δz in equations (4.13) are replaced by a characteristic length. The Péclet number plays an important role for the physical characterization of transport phenomena.

Jean Claude Eugene Péclet (1793-1857), professor in Marseille and Paris, published his famous book '*Traité de la Chaleur et de Ses Applications*' in 1829.

enough. In comparison with equation (4.7) they can be stated as follows: choose the grid-spacing so fine that numerical dispersion is smaller than real physical dispersion.

Conditions (4.14) can be found in mathematical literature criteria as stability criteria for the CIS discretization (4.5) (Meis/Marcowitz 1978). In practice, results using CIS show oscillations near sharp fronts even if the grid-Péclet numbers are slightly below 2.

On the other hand, the BIS-discretization equation (4.4) is reported to be unconditionally stable. But it turns out in practical use that stability is paid for by introducing numerical dispersion. Correction schemes overcome the obstacle, but their use is justified only if the grid-Péclet-criteria are fulfilled. Otherwise the coefficients of second order terms become negative!

Finally, the discretization rules for first and second order terms of the differential equation lead to linear systems of equations. These are sparse, i.e. in each row or column there are only few nonzero elements. In rectangular grids the nonzero elements in the matrices show a diagonal structure. If the canonical numbering is used (first x-, then z-direction), systems with five or nine diagonals appear: the number of diagonals is equal to the number of points in the stencil.

Discretization of the flow equation always leads to a symmetrical matrix with five diagonals. Discretization of the transport yields five diagonals when there is no dispersion, i.e. when Fick's law is valid with a scalar diffusivity constant D. The linear system is symmetrical only in static no-flow cases, which are relevant only for software testing.

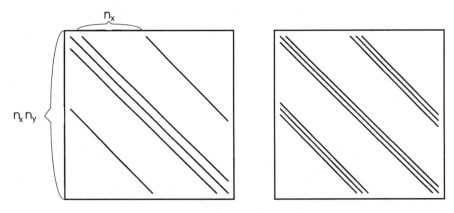

Fig. 4.5 : Diagonal matrices for 5-point and 9-point stencils
n_x denotes number of blocks in direction of preference counting, n_y number of blocks in the other direction

4.2 Temporal Discretization

In a transient simulation the program proceeds along the time axis in timesteps. The same algorithm is used to step from one time level to the other.

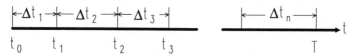

Fig. 4.6: Transient simulation in timesteps

FAST codes organize time-stepping in time-periods. Within each time-period the timestep is constant or calculated automatically by the program - the user has to choose between those two possibilities. The following figure shows the input box in the user shell.

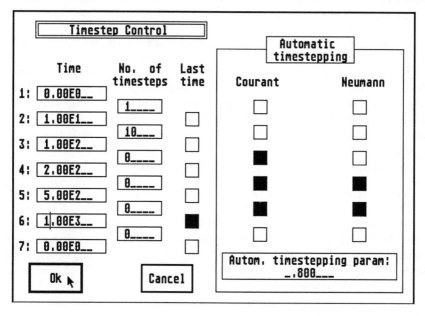

Fig. 4.7: Input-box for timestep selection (ATARI/GEM-version)

The procedure implemented in the FAST codes is a generalized Crank-Nicolson method. The only time derivative in the simplified set of equations appears in the transport differential equation. If the spatial discretization in the transport equation is denoted by Ω, the final discretization (for space and time) can be written as:

$$\frac{\theta(t+\Delta t)-\theta(t)}{\Delta t}=(1-\kappa)\Omega(\theta(t+\Delta t))+\kappa\Omega(\theta(t)) \qquad (4.15)$$

κ is a timelevel weighting factor chosen by the user out of the interval [0,1]. If it is chosen as 1, an explicit algorithm is obtained which is known as forward *Euler* method. Sometimes the notation is simply: forward in time (FIT). All other choices result in an implicit algorithm. The classical *Crank-Nicolson* method is given by equally weighting with $\kappa=\frac{1}{2}$ (central in time, CIT) . The other extreme, where $\kappa=0$ is chosen, leads to the totally implicit algorithm (backward Euler method, backward in time or BIT).

As in most other finite difference implementations, each equation in formulation (4.15) is multiplied with the area measure $\Delta x \cdot \Delta z$. The operator Ω can often be written as

$$\mathbf{W} \cdot \mathbf{p} + \mathbf{y} \qquad \text{with } \mathbf{p} = \left(\theta_1, \ldots \theta_N\right)^T \qquad (4.16)$$

where $\mathbf{p}$ is the vector of unknown values to be calculated. Note that $\mathbf{W}$ can be understood as a matrix which has to be multiplied with the following vector. $\mathbf{y}$ is a vector as well, determined by boundary conditions and source/sink-rates. Then the following results:

$$\mathbf{I}\left(\mathbf{p}(t+\Delta t) - \mathbf{p}(t)\right) = \Delta t\left\{(1-\kappa)\mathbf{W}_{t+\Delta t}(\mathbf{p}(t+\Delta t)) + \kappa\mathbf{W}_t(\mathbf{p}(t)) + \mathbf{y}\right\} \qquad (4.17)$$

The index of $\mathbf{W}$ is introduced here to indicate that the discretization matrix depends on the time level as well. It has also been assumed that $\mathbf{y}$ does not change temporally. This is fulfilled if boundary conditions and sources/sinks are constant in the simulated time period. The FAST-C(2D) code uses this assumption. $\mathbf{I}$ is the matrix that contains the block volume in the main diagonal and zeroes in all other diagonals.

Solving the vector equation (4.17) for the unknown $\mathbf{p}(t+\Delta t)$ gives:

$$\left[\mathbf{I} - \Delta t(1-\kappa)\mathbf{W}_{t+\Delta t}\right]\mathbf{p}(t+\Delta t) = \left[\mathbf{I} + \Delta t\kappa\mathbf{W}_t\right]\mathbf{p}(t) + \Delta t\mathbf{y} \qquad (4.18)$$

This equation is implemented in the FAST-C(2D) code to solve the transport equation. Each time step requires the matrix in the brackets on the left side to be evaluated. The right side is evaluated at the old time level and thus stays constant in the Picard iteration (see chapter 4.5.2). The linear system (4.18) has to be solved for each time step and in each step of the Picard iteration.

The given timestepping algorithm is bound by stability conditions as well. These are formulated in terms of Courant numbers and Neumann numbers:

$$Cou_x = \frac{v_x \Delta t}{\varphi R \Delta x} \qquad\qquad Cou_z = \frac{v_z \Delta t}{\varphi R \Delta z} \qquad (4.19)$$

$$Neu = D\left(\frac{1}{\Delta x^2} + \frac{1}{\Delta z^2}\right)\Delta t \qquad (4.20)$$

The criteria are:

$$Cou_x \leq 1 \qquad\qquad Cou_z \leq 1 \qquad\qquad (4.21)$$

$$Neu \leq \frac{1}{2(2\kappa - 1)} \qquad\qquad (4.22)$$

If a grid is specified by the user, both criteria are conditions for the timestep to be sufficiently small. In mathematical literature it is shown that both conditions are necessary for stability - if $\kappa > \frac{1}{2}$ (Meis/Marcowitz 1978). Proofs mostly treat special cases with regular or even quadratic grids only. It is recommended that the criteria (4.21) and (4.22) should not be strongly violated, even for model runs with $\kappa \leq \frac{1}{2}$.

FAST-C(2D) lets an automatic timestepping be chosen. This option takes the stability criteria into account. Timesteps calculated by the computer may fulfill the Courant-criteria (4.21), the Neumann-criterion (4.22) or both (Fig. 4.7).

A timestepping parameter χ can also be specified. χ determines to what degree the most restrictive of the chosen criteria is fulfilled (or violated). When $\chi=1$ is selected the timestep fulfills all criteria everywhere and the equality in (4.21) and (4.22) holds at least at one location. Table 4.1 provides an overview of the criteria fulfilled with the choice of χ- if desired by the user.

Parameter	Cou_x	Cou_z	Neu
χ	$\leq \chi$	$\leq \chi$	$\leq \chi/2(1-2\kappa)$

Table 4.1: Criteria fulfilled by optional automatic timestepping

4.3 Boundary Conditions

Conditions have to be specified at the boundaries of the model region, for otherwise the problem is mathematically underdetermined, i.e. unique solutions do not exist.

One boundary condition needs to be predefined for each variable at all those block edges, which are boundaries of the model. In block-centered grids, as they are used in FAST-C(2D), the conditions are specified for the medium points of the block edges - in 2D for center points of lines. A condition is homogeneous on a model boundary if the same condition holds for all edges on that boundary.

Two boundary conditions have to be given by the modeler at all block edges of the boundary: one for the flow and one for the transport variable. The FAST-C(2D) graphical user interface has two bars at the model boundaries. These indicate the type of boundary conditions. The inner bars represent conditions for flow and the outer bar for transport (Fig. 4.8).

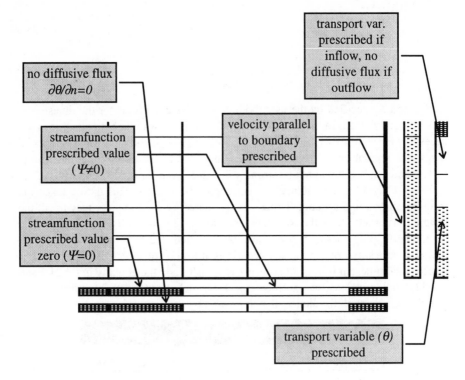

Fig. 4.8: Boundary conditions in FAST-C(2D) (section from GUI-screen)

There are two types of boundary conditions which can be specified by the user when modeling with FAST-C(2D):   • 1st order or Dirichlet - type condition

             or: • 2nd order or Neumann - type condition

If a 1. order type boundary condition is used, the value of the unknown variable is predefined. If a 2nd order type is used, the derivative of the variable is predefined. The following table gives an overview of the physical meaning of boundary conditions.

Type	Flow	Transport
1st order (Dirichlet)	Specified streamfunction	Specified temperature or concentration
2nd order (Neumann)	Specified velocity parallel to boundary	Specified derivative normal to the edge

Table 4.2: Overview on boundary conditions

The no-flow situation is very common and requires the following conditions to be chosen. There is a Dirichlet-type boundary condition for Ψ with the same value along closed edges. A special Neumann-type condition is to be selected for the transport variable: $\partial\theta/\partial n=0$; which means that diffusive or dispersive fluxes across that edge are zero. When the flow boundary condition is of no-flow type, the advective heat or mass flux disappears as well.

There is a restriction in the use of FAST-C(2D) not mentioned in the general overview: the Neumann-condition for transport is always $\partial\theta/\partial n=0$. In fact the derivative of the transport variable is seldom known in practical applications; the generalization $\partial\theta/\partial n\neq0$ is necessary only in few exceptional cases.

On the other hand in FAST-C(2D) does take into account a special mixed type condition. In the condition it has to be distinguished, if there is inflow or outflow at a boundary block edge. In the case of inflow a Dirichlet-type condition with a specified temperature or concentration is used. In the case of outflow the no-flow condition becomes relevant. Note that at some boundaries inflow and outflow at block edges are determined during the model run; thus the boundary condition may change during the simulation run. In other words: the selection of the mixed type hinders the (beforehand) determination, if a Dirichlet- or Neumann-type condition is prescribed.

In the graphical user shell the mixed type boundary condition is denoted as 'inflow Dirichlet type'. It is a useful condition at those edges where the inflow/outflow characteristic is not known beforehand. The mixed type condition is a special 3rd order type boundary condition. General 3rd order or Cauchy-type conditions are not yet implemented in the code.

4.4 Initial Conditions and RESTART

All models require an initial condition to be known and specified by the user. In transient simulations, the specified state for the variables really represents an initial starting point. This is often a simple situation, where density differences are not yet present. One example is a cavity in a no-flow, hydrostatic, isothermal state with no differences in fluid chemistry. Such a situation can be described analytically and then be transformed into a numerical formulation. It can be set up in a laboratory, but it will never be found in field situations.

A simulated laboratory experiment takes the ideal simple situation described above as an initial state for the numerical model. It is much more difficult to find a reasonable initial situation for a field site.

Note that in steady state models the specified input variables are just starting points for the iterative solvers. They do not necessarily have a physical significance. This means it is not necessary that the boundary conditions be fulfilled.

The solver converges faster when a good initial guess is used. But it does not usually make sense, to spend much time getting a better starting point. The program updates the solution anyway and does that quite fast.

A change of initial states can have a huge effect if there are problems with convergence. A divergent could be made convergent with a convenient starting state.

The RESTART-option implemented in most codes allows the continuation of a model run. Two things need to be done:

- in a first run the storage of final state, i.e. the arrays of state variables, is required

- the continuation run requires stored values to be read as initial state

A restart run should be started in a transient model, if the modeler wants to extend the simulated time period. An example: if there are doubts that a steady state has been reached at the end of the first run, RESTART allows to check, if there are significant changes of the variables in the further run.

Temporal changes are not foreseen for all parameters in the program design. They can be simulated using the RESTART-option. An example: if some parameter included in the Rayleigh number undergoes a sudden change at a certain point in time, the simulation can be restarted with a new input value for Ra.

The application of RESTART is apparent in transient simulations, but it does make sense under certain circumstances in steady-state models as well. Two examples:

- doubts about the accuracy of the variables obtained in a first model run, may be checked by a continuation run with a lower value for the accuracy parameters (see chapters 4.5.2 and 4.6)

- if the numerical solver for a modeled situation is divergent, a RESTART-run with a better initial value, obtained in a transient simulation, may converge

4.5 Solution of the Nonlinear System

The discretization of the coupled flow and transport equations generally leads to a set of nonlinear equations. If there is one transport equation and N denotes the number of blocks, volumes, or nodes, the number of equations is $2N$. The whole system is nonlinear, because the discretization of the flow equation depends on the transport variable θ and vice versa.

With appropriate operators D_1, D_2, I_1, I_2, d_1 and d_2 a coupled system of the following type appears as an extension of (4.18):

$$f_1(\Psi,\theta) = \left(\frac{I_1}{\Delta t} - D_1\right)(\Psi) - d_1 = 0$$

$$f_2(\Psi,\theta) = \left(\frac{I_2}{\Delta t} - D_2\right)(\theta) - d_2 = 0$$

(4.23)

There are different methods to solve nonlinear equations. All computer codes for general applications use iterative methods. Two approaches to perform the iteration are generally found in scientific literature: the Newton method which works on the entire nonlinear system, and the Picard iterations which treats parts separately as linear systems.

The ideas which lead to common algorithms are explained in the following. Note that in the FAST-C(2D) code the Picard iterations are implemented. For the equations, solved by FAST-C(2D), the method is equivalent to a variation of the Newton method (see below).

4.5.1 Newton Method and Variations

The Newton method uses the truncated Taylor-series of the functions f_1 and f_2 at location (Ψ^+,θ^+) to derive the iteration algorithm:

$$\begin{pmatrix} \dfrac{\partial f_1}{\partial \Psi} & \dfrac{\partial f_1}{\partial \theta} \\ \dfrac{\partial f_2}{\partial \Psi} & \dfrac{\partial f_2}{\partial \theta} \end{pmatrix} \begin{pmatrix} \Psi^{++} - \Psi^+ \\ \theta^{++} - \theta^+ \end{pmatrix} = - \begin{pmatrix} f_1(\Psi^+,\theta^+) \\ f_2(\Psi^+,\theta^+) \end{pmatrix}$$

(4.24)

The symbols Ψ^{++} and θ^{++} denote the values of the unknown variables after the iteration, while Ψ^+ and θ^+ are start values for the iteration. Equations (4.24) are solved for Ψ^{++} and θ^{++}. Before starting the next iteration step set: $\Psi^+ = \Psi^{++}$ and $\theta^+ = \theta^{++}$.

It is shown in mathematical literature that the Newton method delivers quadratic convergence. From a theoretical point of view the Newton method converges faster than all other methods presented here, but this method has some disadvantages:

- numerical calculations in each iteration are generally quite high
- there is convergence only for relatively good starting points,
              if the start vector is not good, there is divergence
- programming work for implementation is quite high
- derivatives of parameters are needed to implement the method

A simpler variation of the Newton method is found when only the main diagonals are considered within the Jacobi-matrix on the left side of eq.s (4.24). Paniconi/Putti (1993) call this approach the *incomplete Newton method*. The resulting algorithm converges worse than the original Newton method. On the other hand the incomplete Newton method is easier to implement. Note that in each iteration step 2 systems with N unknown values need to be solved and not one big system with $2N$ unknowns.

The incomplete Newton method considers only the nonlinearities within both functions f_1 and f_2. The nonlinearities which arise out of the coupling of the equations are not considered. If both flow and transport equation are linear, the incomplete Newton method is not different from the Picard iterations. Another variation of the Newton method can be constructed in which only one of the off-diagonal-terms of the Jacobi-matrix is neglected.

All methods mentioned above evaluate the right side of equation (4.24) using the most current value for the unknowns, i.e. using the outcome of the previous iteration for the new time level. Note that the Taylor-series is evaluated at (Ψ^+, θ^+). If it is taken at the values from the old time level (Ψ, θ) instead, the *chord-slope* method results:

$$
\begin{pmatrix} \dfrac{\partial f_1}{\partial \Psi} & \dfrac{\partial f_1}{\partial \theta} \\ \dfrac{\partial f_2}{\partial \Psi} & \dfrac{\partial f_2}{\partial \theta} \end{pmatrix} \begin{pmatrix} \Psi^{++} - \Psi \\ \theta^{++} - \theta \end{pmatrix} = - \begin{pmatrix} f_1(\Psi_1, \theta) \\ f_2(\Psi, \theta) \end{pmatrix} \tag{4.25}
$$

The right side is computed with the values at the old time level and does not have to be evaluated in each iteration step. As a result the execution time for one iteration step will be smaller than for the other Newton method variations -except from the first iteration.

The elements of the matrix in (4.25) are determined as chord slopes - this explains the name of the method. Only the evaluation of the functions f_1 and f_2 need to be computed - and not derivatives. This is the main advantage of the algorithm and it is recommended for those cases in which derivatives are difficult to determine (Huyacorn/Pinder 1983).

Of course the reduction of time for single iterations only pays off, if the number of iterations needed to reach a predefined accuracy is not increased too much. The special nonlinearity determines, if the application of a numerical method requires greater or lesser computational work. No general rule can be given.

4.5.2 Picard Iterations

Picard iterations solve the first equation of (4.24) for Ψ and the second for θ. Coefficient functions are evaluated at the old time level. It holds:

$$\left(\frac{I_1(\Psi^+,\theta^+)}{\Delta t} - D_1(\Psi^+,\theta^+)\right)(\Psi^{++}) - d_1(\Psi^+,\theta^+) = 0$$

$$\left(\frac{I_2(\Psi^{++},\theta^+)}{\Delta t} - D_2(\Psi^{++},\theta^+)\right)(\theta^{++}) - d_2(\Psi^{++},\theta^+) = 0$$

(4.26)

Picard iterations have advantages compared with the Newton method:

- iterations are relatively cheap - in terms of computer resources.
- the method converges even with bad starting points
- they are easy to implement
- no derivatives of functions need to be evaluated

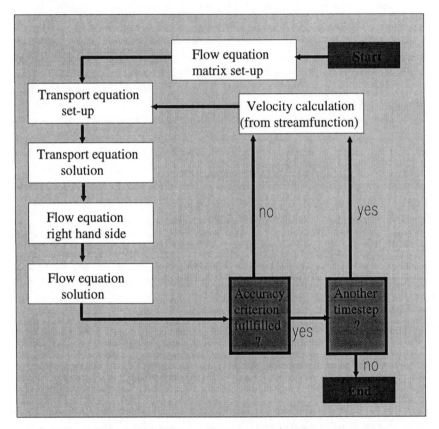

Fig. 4.9: FAST-C(2D) general structure of transient simulations

Experience shows that the Newton method converges faster - if it converges at all. The disadvantage of divergence from inconvenient starting points can be overcome if the following procedure is applied: a limited number of Picard iterations are performed before the Newton method iteration is used. This combination of both procedures is generally recommended.

Note that Picard iterations and the incomplete Newton method, as introduced above, are identical if both equations f_1 and f_2 are linear. The off-diagonal terms in matrix (4.24) then vanish anyway.

As mentioned above FAST-C(2D) allows the use of the Picard iterations only. Both systems, derived from the flow and the transport differential equation (for example: (3.11) and (3.18)), are linear. The only nonlinearity to be tackled by FAST-C(2D) is introduced into the problem by the coupling of both equations. In the FAST-C(2D) code an inner iteration procedure is implemented to solve the nonlinearity. This approach is similar to Picard iterations.

The discretized transport equation is solved first, using an initial flow field. This is a linear equation, and one of the methods described in chapter 3.4 is used for the solution. The solution for the transport variable θ updates the right side of the flow equation is updated - θ does not occur on the left side of the discretized equation for the streamfunction. After the linear system for Ψ is solved, in the next step the new flow field serves for a new simulation of the transport step. To do that, velocity components have to be determined from streamfunction values using equations (4.8) and (4.9).

This loop is performed until a user-defined accuracy criterion is fulfilled. The end of the inner iteration requires that the maximum relative change of both variables is small or that a maximum number of iterations has already been reached.

Fig. 4.9 illustrates the general structure of the loop. The *Picard inner iteration* lies within the timestepping loop. Nevertheless, within the inner iteration further iterations are performed to solve linear systems.

Steady state models require the user to specify start arrays for the flow and transport variable, needed to begin the iteration. Transient simulations use the values specified by the modeler in the first time step: then initial and start values are identical. Further timesteps take the solutions of the old time level to start the Picard iteration loop.

4.6 Solution of Linear Systems

Independent of the method of discretization - whether *Finite Elements, Finite Differences* or *Finite Volumes* or *Method of Characteristics* - the procedure leads to linear systems. In the inner kernel of all codes a solution method for linear solvers is implemented. Fast performing linear solvers are the main components of a fast code.

General linear systems can be written in matrix-vector form:

$$\begin{pmatrix} a_{11} & a_{12} & \cdots & a_{1N} \\ a_{21} & a_{22} & \cdots & a_{2N} \\ \cdots & \cdots & \cdots & \cdots \\ a_{N1} & a_{N2} & \cdots & a_{NN} \end{pmatrix} \cdot \begin{pmatrix} x_1 \\ x_2 \\ \cdots \\ x_N \end{pmatrix} = \begin{pmatrix} b_1 \\ b_2 \\ \cdots \\ b_N \end{pmatrix} \tag{4.27}$$

Systems derived from the discretization of partial differential equations always have few non-zero entries in each line and in each column of the matrix: numerical mathematics names such matrices *sparse*.

There are special numerical methods to treat sparse systems. Only those matrix entries are stored which may be non-zero. This leads to a tremendous saving of computer storage. Further improvement concerning computer resources is given by algorithms especially designed for the solution of sparse systems.

For unstructured grids the matrix becomes unstructured. The irregular structure of - for example - triangular elements is not the only factor which determines the distribution of non-zero entries in the matrix. The numerical order of the nodes also has a great influence on the bandwidth of the matrix. The solution method benefits from a small bandwidth. A variety of algorithms have been invented to find the optimal numerical order.

Structured grids with canonical numerical order (count x first, then y, then z) lead to banded matrices: non-zero elements are placed in diagonal bands parallel to the main band from top left to the lower right corner of the matrix. The diagonals directly adjacent represent two neighboring blocks in x-direction.

Fig. 4.10 shows the case for rectangular grids. The 1D-case has three bands: the main diagonal and the directly adjacent ones. The 2D-case has two additional diagonals in distance n_x from the main diagonal. These form the whole system in the case of 5-point stencils. There are four more diagonals around the outer ones if a 9-point stencil is used (see Fig. 4.4).

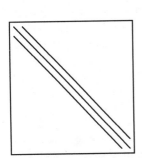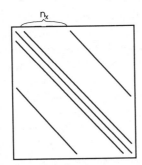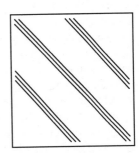

Fig. 4.10: Banded matrix-structure for 3-point-, 5-point- and 9-point-stencils in 2D (only those diagonals are shown, which possess possibly non-zero entries)

The formulation in a rectangular grid has the advantage that non-zero values of matrix elements can occur in certain bands. There is no need to determine and store the position of non-zero entries. The solution algorithm can be easy and fast. In irregular grids the simple structure of the matrix cannot generally be maintained.

The solvers to be used become more elaborate and slow. On other the hand, the number of unknowns can be reduced if a complex model region is appropriately represented by a grid of irregular triangles.

The linear systems arising from the discretization of flow and transport differential equations are generally non-symmetrical. Nevertheless for flow and diffusion problems special transformations lead to symmetrical form. *Finite differences* details are described by Holzbecher (1996). A similar procedure may work for *finite elements*. The technique of *finite volumes* brings the matrices directly in symmetric form. This is an important point for the performance of the model, because solvers for symmetric matrices converge much better than those for non-symmetric systems. The number of elements to be stored is also smaller.

The linear systems arising from a transport equation with non-zero advection are always non-symmetric. A special procedure proposed by Leismann/Frind (1989) leads to symmetric form, but is restricted by severe stability criteria.

Solvers for linear systems can be either *direct* or *iterative*. Direct solvers, like the *Gauß*-elimination method, approach the solution directly. Iterative solvers use a certain iteration specification which - starting from an initial guess applied again and again - improves the solution in the case of convergence. For simple cases the improvement of the solution may be given in each step, but it is often possible that the quality of the solution may become worse in certain steps.

Iterative algorithms are almost always preferred over direct methods nowadays. The main reason is that direct methods have the disadvantage of changing the matrix in the course of calculation. Positions with zero element in the original matrix may get a non-zero entry. This point is crucial for large sparse linear systems as they emerge from the discretization of flow and transport problems. Storage requirements for the computer generally become too high. An example: if the *Gauß* method is applied on the systems as shown in Fig. 4.10, the matrix will be filled within the two outer bands.

Iterative methods of the first generation are the *Gauß-Seidel-* and the *Jacobi-*algorithms. Several improvements have been found since then. Conjugate gradient (CG) methods are direct methods in the concept of the inventors: the exact solution should be reached in N steps, if the number of unknown is N. Computer operations with a finite number of digits make it impossible to reach the exact solution. Programmers use conjugate gradients as iterative solvers (see below). Another iterative procedure favored in recent software is the multigrid method.

In an iterative algorithm, a given approximation of the solution vector x_0 is improved in steps following a special routine. In one iteration, the approximation x_i is changed to give a new approximation x_{i+1}. The algorithm stops if one of the following criteria are fulfilled:

$$\frac{\|x_{i+1} - x_i\|}{\|x_{i+1}\|} \leq \varepsilon \quad \text{or} \quad \frac{\|Ax_{i+1} - b\|}{\|x_{i+1}\|} \leq \varepsilon \quad (4.28)$$

An endless loop is avoided by specifying as an additional criterion for the end of the iterations.

Various reasons may cause the maximum number of iterations to be performed in the solution algorithm. Maybe the algorithm diverges, i.e. it does not approach a solution. Maybe the required accuracy is too low. The modeler can check which alternative is true by changing the accuracy and/or the maximum number of iterations.

If the modeler sets accuracy too low, the algorithm ends with the maximum number of iterations, even if the method converges.

If accuracy is not low enough, the variables calculated at the end of the iteration loop are far from the required solution. Note that the specified value for accuracy is not the maximum difference between numerical and analytical solutions. Depending on the option chosen in the menu, the accuracy at the end of a usual iteration loop is

- the maximum distance between start and final state of the variables for the last iteration
- the maximum value of the residual, i.e. $\left\| A\mathbf{x}_{i+1} - b \right\|$

Accuracy criteria can use different *norms* to define the distance between solution vectors. FAST-programs offer an option between two possibilities; the maximum-norm and the 2-norm:

$$\left\| \mathbf{x} \right\|_{max} = \max\left\{ \left| x_i \right|, i = 1,...N \right\} \qquad \left\| \mathbf{x} \right\|_2 = \sqrt{\sum_{i=1}^{N} x_i} \qquad (4.29)$$

4.6.1 Conjugate Gradients

The method of conjugate gradients is a fast solver for large sparse linear systems

$$\mathbf{A} \cdot \mathbf{x} = \mathbf{b} \qquad (4.30)$$

where $\mathbf{x}$ and $\mathbf{b}$ are vectors of length N and $\mathbf{A}$ is a $N{\times}N$ matrix. The idea of the method is to start from $\mathbf{x_i}$ to search a new approximation $\mathbf{x_{i+1}}$ in a certain direction $\mathbf{p_i}$:

$$\mathbf{x_{i+1}} = \mathbf{x_i} + \alpha_i \mathbf{p_i} \qquad (4.31)$$

The scalar α_i is determined in such a way that the functional

$$F(\mathbf{y}) \equiv \frac{1}{2} \mathbf{y}^T \mathbf{A} \mathbf{y} - \mathbf{y}^T \mathbf{b} \qquad (4.32)$$

is minimized for all vectors in search direction. Note that the minimum of the functional in the N-dimensional vectorspace is the solution of the linear system. For an iteration the search direction is determined as follows:

$$\mathbf{p}_{i+1} = \mathbf{b} - \mathbf{A}\mathbf{x}_{i+1} + \beta_i \mathbf{p}_i \qquad (4.33)$$

where $\mathbf{p}_i$ and $\mathbf{p}_{i+1}$ are conjugated. It can be shown that $\mathbf{x}_i$, the ith iterate, minimizes the functional on the subspace $\left\{ \mathbf{r}_0 = \mathbf{A}\mathbf{x}_0 - \mathbf{b}, \mathbf{A}\mathbf{r}_0, ..., \mathbf{A}^{i-1}\mathbf{r}_0 \right\}$, the so called *Krylov* subspace exactly (for that reason the CG can be found also under the name *Krylov subspace methods*). Theoretically after N steps the exact solution of the linear system is found. At first sight CG seems to be a direct method. Due to round-off errors and the limited representation of real numbers on a computer, the algorithm generally does not yield the exact solution - even if N iteration steps are performed.

The required accuracy ε is usually reached with much fewer iterations. The conjugate gradient method is thus used as an iterative method. In PASCAL-notation the classical algorithm writes as follows:

$$\mathbf{r}_0 := \mathbf{A}\mathbf{x}_0 - \mathbf{b}; \quad \mathbf{p}_{-1} := 0; \quad \varsigma_{-1} := 1; \quad i = 0;$$

while   *residuum > tolerance*   do

begin

$$\varsigma_i := \mathbf{r}_i^T \mathbf{r}_i; \quad \beta_i := \varsigma_i / \varsigma_{i-1};$$
$$\mathbf{p}_i := \mathbf{r}_i + \beta_i \mathbf{p}_{i-1};$$
$$\mathbf{z}_i := \mathbf{A}\mathbf{p}_i; \qquad\qquad\qquad (4.34)$$
$$\sigma_i := \mathbf{p}_i^T \mathbf{z}_i; \quad \alpha_i := \varsigma_i / \sigma_i;$$
$$\mathbf{r}_{i+1} := \mathbf{r}_i - \alpha_i \mathbf{z}_i;$$
$$\mathbf{x}_{i+1} := \mathbf{x}_i - \alpha_i \mathbf{p}_i;$$
$$i := i + 1$$

end;

Bold letters in (4.34) denote vectors, while scalar parameters are represented by normal Greek letters. The new search direction $\mathbf{p}_i$ is calculated as a combination from the old search direction $\mathbf{p}_{i-1}$ and the residual

$$\mathbf{r}_i = \mathbf{A}\mathbf{x}_i - \mathbf{b} \qquad (4.35)$$

It can be shown that the given formula for α_i gives the exact position of the minimum of the function F of (4.32) in the search direction. Note that the residual is determined iteratively as well - not by using (4.35). A high number of iterations may cause additional errors. The method implemented in FAST-C(2D) checks the residual again before leaving the solver module. If it turns out that the required accuracy is not reached, the iterative procedure is started again (post-iterations).

The classical CG-algorithm has been invented for matrices **A**, which are symmetrical and positive definite. The convergence of the method can be proved for this type of matrices (Hestenes 1980). Nevertheless, the algorithm as described can be applied for non-symmetric, non positive definite matrices as well. Practice shows that CG in its classical form may converge and deliver reasonable results.

Variations of the classical algorithm have been developed for non-symmetric matrices. Most important are: BI-CG (Fletcher 1976), CGS (Sonneveld 1984), BI-CGstab (van der Vorst 1990) and ORTHOMIN (Vinsome 1976). In FAST-C(2D), the first three variations are available for solution of the - generally non-symmetric - linear systems, which are derived from the transport equation.

The speed of convergence on the computer is improved by applying so called *preconditioners*. The most frequently used preconditioners are the incomplete *Cholesky*-method (Schwarz 1984) and the overrelaxation method (Evans 1967). Multigrid has been recommended recently as preconditioner (Axelsson/Vassilevski 1989). The success of preconditioning lies in the reduction of iterations needed to reach a specified accuracy. On the other hand, the computational work in one iteration is always higher than in the classical form of CG. No general guideline can be given: which method performs best in terms of solver execution time, depends on the specific problem (i.e. matrix and right side **b**) and on the computer type.

When *scaling* or *preconditioning* is used, the matrix **A** of the linear system is transformed into a more appropriate form. A matrix comes in 'scaled' form when columns and rows are multiplied with convenient factors, to get 1.0 elements in the main diagonal only. Besides the fact that CG may converge better on 'scaled' matrices, it has to be considered that fewer mathematical operations have to be performed in the matrix-vector multiplications.

Preconditioning applies a transformation of the following type:

$$\mathbf{A} \to \mathbf{P}_L \mathbf{A} \mathbf{P}_R \qquad (4.36)$$

Matrices $\mathbf{P}_L$ and $\mathbf{P}_R$ need to be chosen appropriately.

Finally, it can be stated that CG-type methods a set of iterative solvers are fast and robust. The first property - speed - comes into focus in comparison with older iterative methods, like *Jacobi-* or *Gauß-Seidel* method including (SOR) (Stoer/Bulirsch 1973). Compared with other modern methods (like multigrid), CG-solvers have the advantage that they can be successfully applied in unmodified form to a broad class of problems.

Modelers using FAST-C(2D) can choose various solver options. For the flow equation they can use CG in the classical and preconditioned form. For the transport equation they have the additional choice between four types of solvers for non-symmetric matrices.

In FAST-C(2D) code SOR-type preconditioning is implemented. It can be controlled by an overrelaxation factor. This factor usually has its optimum value in the interval between 1.0 and 2.0. Below the optimum, the method shows only slightly less performance (Holzbecher 1991a). But it converges much worse if the overrelaxation factor is slightly above the optimum. For 2D-problems the factor is mostly not much different from 1.5.

The user can also choose to run the solver in double-precision or single-precision mode. The single-precision mode, in which float numbers have a mantissa of normal length, is usually sufficient. Results can be checked or improved by using the double-precision mode with float numbers of double length. Note that the execution time for one iteration in double precision mode is longer. It can be observed that the number of iterations in double precision mode is often higher than in single precision mode.

Which accuracy can be reached depends on the precision mode. In single precision mode ε should usually not be smaller than 10^{-6}; in double precision mode not smaller than 10^{-12}. In exceptional cases the iterative solver may reach convergence for 10^{-7} in single, for 10^{-13} in double mode.

The user has additional options concerning the form of accuracy criteria. The norm (distance) can be selected (see (4.29)) and there is a choice between the use of residual or solution change in the criterion (see(4.28)).

4.7 Postprocessing

There are various options for output and postprocessing. The user can look into arrays of the calculated variables after each or after a fixed number of timesteps. Besides the main two variables, velocity component and viscosity arrays can be written on screen or into files optionally.

There is also an option to output eddy centers, Nusselt numbers and extrema of velocities. The following treats the calculation of Nusselt numbers from the numerically θ-fields in some detail. Linear, quadratic or cubic approximation can be used:

$$Nu^{(\eta)} = \frac{1}{L} \sum \frac{\partial \theta}{\partial z}^{(\eta)}$$

(4.37)

with:

$$\frac{\partial \theta}{\partial z}^{(1)} = \frac{2}{\Delta z}(\theta_1 - \theta^*) \qquad \frac{\partial \theta}{\partial z}^{(2)} = \frac{1}{\Delta z}[3(\theta_1 - \theta^*) - \frac{1}{3}(\theta_2 - \theta^*)]$$

$$\frac{\partial \theta}{\partial z}^{(3)} = \frac{1}{\Delta z}[\frac{15}{4}(\theta_1 - \theta^*) - \frac{5}{6}(\theta_2 - \theta^*) + \frac{3}{20}(\theta_3 - \theta^*)]$$

(4.38a,b,c)

Here θ^* denotes the preset value of θ at the boundaries and θ_1, θ_2, θ_3. numerically derived values in the three blocks next to the boundary. The sum in equation (4.37) has to be extended over all blocks on a horizontal boundary. In steady state conditions the calculations were made for upper and lower edge separately to check the computational output. The quadratic approximation is always greater than the linear, and the cubic lies between both lower order approximations.

Graphical representation of variables on the screen is also an additional option. There are several other options concerning the form of the output: lengths and width on the screen, number of contours for each variable, automatic/manual determination of contour-levels, and more. In transient simulation runs, pictures can be drawn on the screen after each time step. Thus there is an automatic visual control on the computer run enabled. The calculation of the results is usually quite fast, the development of a system can be viewed on the display like a animation sequence.

The FAST-C(2D) program can save flow pattern figures as bit-maps. Using an imaging software tool the modeler can gather the saved figures in an animated bitmap which can be viewed by common browsers[4].

---

[4] Examples, showing density driven flow patterns for examples of this book, can be viewed in *FastDemo.HTM* on CD-ROM and in the Internet; see also: *FastDemo.HLP*.

5 Steady Convection

5.1 Bénard Experiments in Porous Medium

At the beginning of the century, H. Bénard (1900) set up experiments with a fluid, whale oil, in a box heated from below. When he gradually heated the box at the bottom, he observed that at a certain marginal temperature the initially motionless system started to move. After a certain time, almost regular stable flow patterns appeared. The convection cells had hexagonal shapes when viewed from the top. Flow in the middle of the hexagons was upward, while flow at the edges of the hexagons was downward.

2D rolls instead of hexagons may appear, if the experiment is performed under different conditions. Fig. 5.1 shows a convective pattern in a 2D cross-section by streamlines. Similar figures have been obtained from experiments using dye or suspended particles. Five convection cells can be distinguished in the figure. The rotation within the cells changes from one to the other: neighbor cells have a different rotation. Thus near the cell boundaries alternating up- and downflow regions can be observed.

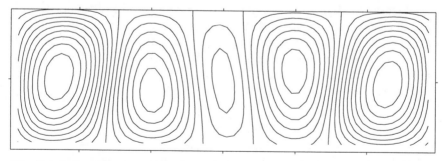

Fig. 5.1: Irregular convective motion (visualized by streamlines - obtained from a numerical simulation in a 2D cross-section with FAST-C(2D)

Experiments similar to those conducted by Bénard have been conducted by many researchers in the 20th century and the outcomes were similar. At temperatures below the critical margin the system remains in the static no-flow situation set up before heating. The temperature in the fluid changes only with depth: lower at the upper boundary, higher at the bottom.

The situation is totally different when the critical margin is exceeded: flow can be observed. There are regions with high temperatures reaching almost to the top, and other regions where low temperatures can be found near the bottom (see Fig. 5.2).

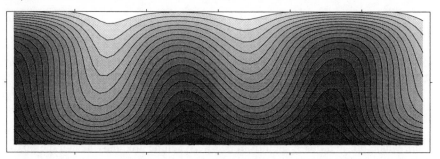

Fig. 5.2: Temperature distribution in the case of convection (the presented
distribution corresponds with the streamline pattern of Fig. 5.1)

Bénard's experiments mark the starting point in the investigation of convective flow patterns. Various patterns have been identified since then: besides 3D-hexagons, there are 2D-rolls as a steady state pattern. Various experimental investigations have been performed in different geometries: 2D-horizontal, 2D-vertical, 3D. The system has been heated from below and cooled from above. Recent experiments focus on unsteady regimes, which may be oscillatory and on routes to chaotic motions.

Bénard attributed the observed phenomenon to density differences. Heat introduces a potentially unstable situation, because higher density fluid lies above lower density fluid. In order to cause the observed effect, the instability needs to be sufficiently strong. In Lord Rayleigh's pioneering work a quantitative characterization for the critical margin could be derived. In his paper, '*On convection currents in a horizontal layer of fluid when the higher temperature is on the under side*', Lord Rayleigh (1916) introduces a number, known today as the *Rayleigh number*. It is defined as:

$$Ra = \frac{g\Delta\rho H^3}{\mu D} \tag{5.1}$$

Besides density gradients, other reasons have been observed, which initiate instabilities in fluid systems. In fact it turned out that the effects observed by Bénard in his pioneering experiments were determined mainly by a temperature dependent surface tension - a so-called Marangoni instability (Oertel 1982) - and not by temperature differences. In that respect it must be noted that the fluid layer in Bénard's experiments had a height of approximately 1 mm and that the box was not closed at the top. Nield (1964) suggests that a combined effect of buoyancy and surface tension may have induced fluid motions in the original Bénard set-up.

A quantitative criterion for the porous medium case was derived by Lapwood (1948), in following the procedure introduced by Lord Rayleigh (1916) for the pure fluid case. The critical margin in most publications is formulated with respect to a dimensionless parameter, the *Rayleigh number,* which has been defined already in formula (3.47).

Publication	Fluid	Solid
Morrison 1947	Glycerol, Carbon Tetrachloride, Water	Unconsolidated sand
Morrison e.a. 1949	"	"
Rogers/Morrison 1950	", Silicone fluids	"
Roers/Schilberg 1950	Water	"
Wooding 1959	Distilled water	Glass spheres
Wooding 1962	Potassiumpermanganate solution	Glass sheets
Wooding 1963	", Sodium sulfate solution	Glass beats
Schneider 1963	Distilled water, Turpentine, Oil, Air	Glass/Steel Spheres/Rings
Katto/Masuoka 1967	Nitrogen gas	Glass, Steel, Aluminum
Elder 1967a	Water	Glass spheres
Elder 1967a, 1967b, 1968	Silicone oil	-
Bories/Thirriot 1968	Silicon oil	Glass beats
Masuoka 1972	Water	Glass balls
Combarnous/Bories 1973	"	"
Kaneko e.a 1974	Heptan, Ethanol	Silica sand
Buretta/Berman 1976	Water	Glass beats
Bau/Torrance 1982	"	", Sand
Close e.a. 1985	Air	Polypropylene Spheres
Jonsson/Cattan 1987	Water, Oil, Mercury	Glass, Steel, Lead

Table 5.1: Overview: experiments on convection in porous media

Bénard-type experiments have been conducted to study convective motions in porous medium. Different types of fluids, liquids and gases, have been used to study flow in different types of porous media, mostly artificial glass products. An overview is given in Table 5.1; the list is probably incomplete. Most experiments have been conducted using heating and cooling facilities at the bottom and the top of the set-up. Only Wooding (1959, 1962a, 1962b) performed experiments with fluids of different salinity. Elder (1967) conducted his experiments in Hele-Shaw cells. Different measurement techniques have been used. Some scientists used suspended solids to visualize flow patterns (Morrison 1947, Morrison e.a. 1949, Rogers/Morrison 1950, Bories/Thirriot 1968) and some used dye (Wooding (1959, 1962a, 1962b). Rogers/Schilberg (1950) took radioactive tracers and a Geiger-Müller counter. Most researchers measure temperatures directly at various locations within the system.

Some studies are mainly directed to the determination of total heat transfer. Schneider (1963) seems to be the first who used the idea of studying mean heat transfer through the system in order to find the onset of convection. In fact, the (mostly electrical) energy needed to keep the boundaries at constant temperatures is proportional to the mean heat flux, if energy losses to the environment can be neglected. The onset of convective motions can be recognized by an increase of mean heat transfer; in other terms: by values of the Nusselt number above 1 (see chapter 3.8).

For pure fluid convection the same technique was first applied by Schmidt/Milverton (1935). Chandrasekhar (1963) lists a number of other experiments for the pure fluid case. The first who adopted the technique for porous medium convection after Schneider (1963), were Katto/Masuoka (1967) and Masuoka (1972). Further references can be found in Table 5.1.

The Bénard case, despite its simplicity concerning geometry and boundary conditions, turns out to be most important for studies of density driven phenomena. Key characteristics, like the Rayleigh and Nusselt numbers are closely connected with studies on Bénard-type experiments.

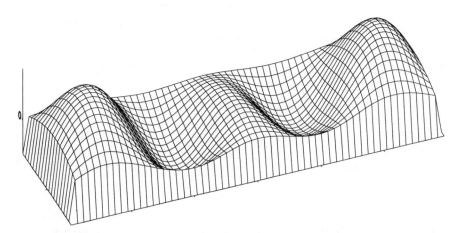

Fig. 5.3: Streamfunction surface of 2D convective motions (for the cross-section shown in Fig. 5.1)

5.2 Linear Analysis

5.2.1 Isotropic Porous Medium

Here I follow Lord Rayleigh's famous argumentation, which he published in 1916 in his paper '*On convection currents in a horizontal layer of fluid when the higher temperature is on the under side*'. In contrast to Lord Rayleigh, the porous medium situation is treated in the following. The analysis of the problem uses the same set of equations, which is the basis for numerical simulations with FAST-C(2D). The equations are derived from basic principles in chapter 3 (more specifically: the dimensionless variables θ and Ψ are used, which are introduced in chapter 3.7). Chapter 4 outlines, how a numerical code is built from these. Fig. 5.3 shows the streamfunction as a surface above the vertical cross-section in which convective motions are described.

A solution requires the equations (3.41) and (3.45) to be fulfilled, in which θ is considered as a dimensionless temperature:

$$\nabla^2 \Psi = -Ra \, \partial\theta \, / \, \partial x$$

$$\nabla^2 \theta - \mathbf{v}\nabla\theta = \frac{\partial\theta}{\partial t}$$

$$(5.2)$$

θ can generally be viewed as a dimensionless transport variable. It can describe either normalized temperature or normalized salinity: only the sign on the right hand side of the flow equation is different. The thermal case obviously requires the fulfillment of the following boundary conditions:

(1) $\theta = 0$ at the top / $\theta = 1$ at the bottom
(2) $\partial\theta/\partial x = 0$ at vertical boundaries
(3) $\Psi = 0$ at all boundaries

The physical explanation for boundary condition (2) is that there is no diffusive/dispersive flux out of the system: the walls are perfectly insulated. There is no advective flux across the vertical edges either - thus (2) is in fact a no-flow boundary condition. Fig. 5.4 gives a schematic illustration of the boundary conditions for the thermal and the saline case in a finite box.

Besides the no-flow characteristic, the static situation shows a linear increase of temperature from the top to the bottom. With dimensionless variables the static state is given by:

$$\theta = z \qquad\qquad \text{and} \qquad\qquad \Psi = 0 \qquad\qquad (5.3)$$

For the saline case holds: $\theta = 1\text{-}z$. The following analysis treats the thermal case, but an analogous treatment can be made for the saline case. The solutions of the nonlinear system are split into two parts. One part represents the static solution (with conduction only) and the other the perturbation. For the moment I denote the perturbation variables with θ^* and Ψ^* (despite the fact that for the latter variable it would be necessary to introduce a new notation). Thus holds:

$$\Psi^* = \Psi \qquad \text{and} \qquad \theta^* = \theta - z \qquad (5.4)$$

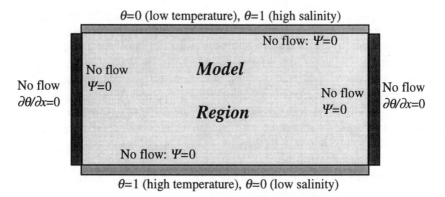

$\theta=0$ (low temperature), $\theta=1$ (high salinity)

No flow: $\Psi=0$

No flow
$\Psi=0$

Model

No flow
$\Psi=0$

No flow
$\partial\theta/\partial x=0$

No flow
$\partial\theta/\partial x=0$

Region

No flow: $\Psi=0$

$\theta=1$ (high temperature), $\theta=0$ (low salinity)

Fig. 5.4: Flow and transport boundary conditions for thermal and saline
        convection in streamfunction formulation (flow condition is depicted
        inside the model region; transport boundary condition is depicted outside)

The characterization of the critical margin makes use of the linearized problem. For the porous medium case, I follow the basic approach which Lord Rayleigh introduced for pure flow (Rayleigh 1916). Setting solution (5.3) and perturbation (5.4) into the equations (3.32) and (3.36) (for dimensionless variables), one gets:

$$\nabla^2 \Psi^* = -Ra\, \partial\theta^* / \partial x$$

$$\nabla^2 \theta^* + \frac{\partial\Psi^*}{\partial z}\frac{\partial\theta^*}{\partial x} - \frac{\partial\Psi^*}{\partial x}\left(\frac{\partial\theta^*}{\partial z}+1\right) = \frac{\partial\theta^*}{\partial t} \qquad (5.5a,b)$$

Most terms of the static part vanish. Only one term remains, when the derivatives are taken. It is: $\partial\theta/\partial z=1$. The next step to reach the linearized problem is to omit terms with quadratic perturbations. The following system remains:

$$\nabla^2 \Psi^* = -Ra\, \partial\theta^* / \partial x$$

$$\nabla^2 \theta^* - \frac{\partial\Psi^*}{\partial x} = \frac{\partial\theta^*}{\partial t} \qquad (5.6a,b)$$

Following Lord Rayleigh's line of argumentation, for the further analysis, I consider the system of infinite length, i.e. boundary conditions for the vertical boundaries can be neglected. The perturbation variables then obviously need to fulfill the slightly different set of boundary conditions:

(1) $\theta^* = 0$ at the top and at the bottom[1]
(2) $\Psi^* = 0$ at all boundaries

Steady state cells are assumed to have the explicit form

$$\Psi^* = \Psi_0 \sin(r\pi x / L)\sin(s\pi z)$$

$$\theta^* = \theta_0 \cos(r\pi x / L)\sin(s\pi z)$$

(5.7)

θ_0 and Ψ_0 are the amplitudes of the mode characterized by the numbers r and s. Note that with respect to the first boundary condition, only integer values for s make sense. The number of cells in horizontal direction in an interval of length L is r, while s is the number of layers of convection cells. The cell aspect ratio ϑ, its height compared to the length, is given by:

$$\vartheta = \frac{H}{s}\frac{r}{L}$$

(5.8)

Lapwood (1948) uses the form $\theta^* = \theta_0 \sin(lx)\sin(s\pi z)$, which is equivalent to the formulation (5.7) for the existence of solutions in the system with infinite horizontal extension - the position relative to vertical axis is different only. An extended form for the functions Ψ^* and θ^* is given in the following chapter. In fact (5.7) is the most simple form of the spectral approach (Bories 1987):

$$\theta^* = \sum_{m=0}^{M}\sum_{s=0}^{S}\theta_{ms}(t)\cos(m\pi x)\sin(s\pi z)$$

(5.9)

with $M=S=1$, $\theta_{00}=0$, $\theta_{10}=0$, $\theta_{01}=0$, $\theta_{11}=\theta_0$ and $m=r/L$ (it is an extension insofar that for m some rational values are allowed additionally).

A condition for the steady state solutions is obtained from equations (5.6a,b) by omitting the time derivative. Further replacement of θ^* and Ψ^* using (5.7) yields a set of two linear equations:

$$\begin{pmatrix} -\pi^2(s^2 + (r^2/L^2)) & r\pi Ra/L \\ r\pi/L & -\pi^2(s^2 + (r^2/L^2)) \end{pmatrix}\begin{pmatrix} \Psi_0 \\ \theta_0 \end{pmatrix} = \begin{pmatrix} 0 \\ 0 \end{pmatrix}$$

(5.10)

The determinant condition that there are nonzero solutions (θ_0, Ψ_0) gives a relationship between Ra, s and r:

$$\pi^4[s^2 + (r^2/L^2)]^2 - Ra(r\pi)^2/L^2 = 0$$

(5.11)

---

[1] Lapwood (1948), following Lord Rayleigh (1916), at this point demands the additional condition that $\partial^2\theta/\partial z^2 = 0$ holds at top and bottom boundaries.

The quadratic equation in r^2 can be solved:

$$r_{1,2} = \frac{L^2}{2\pi}(\sqrt{Ra} \pm \sqrt{Ra - 4\pi^2 s^2})$$

(5.12)

Equation (5.12) delivers real values for r only if: $Ra \geq 4\pi^2 s^2$. The minimal marginal number for all positive integers s is obviously given for $s=1$: $Ra_{crit} = 4\pi^2$. This is the classical result. Then for $s=1$ there exists a possible convective pattern with real valued r. If the Rayleigh number as the characteristic value of a geothermal system is below the critical margin, no real values for r exist: the only solution of the equation is the one which describes the pure conduction state.

If $Ra>4\pi^2$, the conduction state may become unstable. The original papers of Rayleigh (1916) and Lapwood (1948) contained studies of hypothetical systems of infinite length. This geometric condition lets all values of r be realized. This is not true in (real) systems of finite length L. There a finite number of cells of equal size need to fill the entire length. Only when r is a positive integer does a solution with convective flow patterns come into existence.

Equation (5.11) allows critical values to be determined for all values of r, s and L. For given L it can be defined:

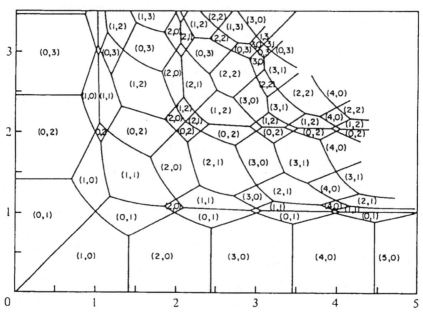

Fig. 5.5: Preferred modes in dependence of side lengths L_1 and L_2 in horizontal directions (after Beck 1972)

$$Ra_{crit}^{r,s} = \left\{ \frac{\pi L}{r} \left[s^2 + \left(\frac{r}{L} \right)^2 \right] \right\}^2$$

(5.13)

Ra_{crit} is the solution of equation (5.11) and (5.13) for $s=1$ and $r=L$, i.e. $\vartheta=1$. Near $Ra=4\pi^2$ there will be one layer of rolls and, if the length of the box is a multiple of its height, the cells will be quadratic. If L is not an integer the ratio r/L will be different from unity for all values of r and s and the lowest critical Rayleigh number will be different from and slightly higher than $4\pi^2$.

More generally, Beck (1972) studies the mode with minimum Rayleigh number for given side lengths in a 3D-box. Fig. 5.5 reproduces the 2D-graph of the main quadrant of the coordinate systems showing the preferred modes for 3D-boxes with horizontal side lengths L_1 and L_2. Similarly, a graph can be plotted showing the critical Rayleigh-numbers. The intervals on the horizontal axis represent the 2D-roll patterns. The length-interval, where the $(r,0)$-mode is the mode with the lowest Rayleigh number, is marked by the two border-points $\sqrt{r(r-1)}$ and $\sqrt{r(r+1)}$. Riley/Winters (1989) denote these points as modal exchange points and these are associated with a characteristic value of the Rayleigh number which is:

$$Ra = \pi^2 \frac{(2r+1)^2}{r(r+1)}$$

(5.14)

Nevertheless, if the transient development of the system (5.5) is simulated, one will find that these points do not necessarily determine the switch from one mode to another. The system stays on a solution branch chosen before even if those *modal exchange*-points are passed (see chapter 5.4).

5.2.2 Anisotropic Porous Medium

In the anisotropic situation the equations (5.2) have to be replaced by

$$\left(\frac{\partial^2}{\partial x^2} + \frac{k_z}{k_x} \frac{\partial^2}{\partial z^2} \right) \Psi = -Ra^* \, \partial\theta / \partial x$$

$$\left(\frac{\partial^2}{\partial x^2} + \frac{D_z}{D_x} \frac{\partial^2}{\partial z^2} \right) \theta - v\nabla\theta = \partial\theta / \partial t$$

(5.15)

Note that the Rayleigh number is defined using the z-direction permeability and the x-direction diffusivity:

$$Ra^* = \frac{k_z g \Delta\rho H}{\mu D_x}$$

(5.16)

Instead of (5.11) one obtains here

$$\pi^4[\alpha_F s^2 + (r^2/L^2)][\alpha_T s^2 + (r^2/L^2)] - Ra^*((r\pi)^2/L^2) = 0 \qquad (5.17)$$

with $\alpha_F = k_z/k_x$ and $\alpha_T = D_z/D_x{}^2$. For given r and s, again, the critical margin as a generalization of equation (5.13) is characterized by the relation:

$$Ra^{*r,s}_{crit} = \left(\frac{\pi L}{r}\right)^2 \left[\alpha_F s^2 + \left(\frac{r}{L}\right)^2\right]\left[\alpha_T s^2 + \left(\frac{r}{L}\right)^2\right] \qquad (5.18)$$

The determinant condition in this case leads to the inequality

$$Ra^* \geq \left(\sqrt{\alpha_F} + \sqrt{\alpha_T}\right)^2 \pi^2 \qquad (5.19)$$

A mixed definition for the Rayleigh number is given by equation (5.16): for permeability the vertical, for diffusivity the horizontal component appears. Instead the condition (5.19) can be reformulated for Rayleigh numbers, which are defined for permeability and diffusivity both in horizontal direction Ra^x or both in vertical direction Ra^z. Then the convection criterion (5.19) transforms into:

$$Ra^x \geq \left(\alpha_F + \sqrt{\alpha_T \alpha_F}\right)^2 \pi^2 \quad \text{resp.} \quad Ra^z \geq \left(1 + \sqrt{\alpha_F/\alpha_T}\right)^2 \pi^2 \quad (5.20)$$

In order to obtain the aspect ratio ϑ at the onset of convection, the minimum has to be found from the equation

$$Ra^*(\vartheta) = \frac{1}{\vartheta^2}(\alpha_F + \vartheta^2)(\alpha_T + \vartheta^2) \qquad (5.21)$$

where the cell aspect ratio (in the cell with unit height) is defined as $\vartheta = r/L$ - without loss of generality is assumed: $s = H = 1$ (compare (5.8)). For the minimum, the following condition is necessary (setting the derivative with respect to ϑ^2 to zero):

$$-\frac{\alpha_F \alpha_T}{(\vartheta^2)^2} + 1 = 0 \qquad \text{resp.} \qquad \vartheta^2 = \sqrt{\alpha_F \alpha_T} \qquad (5.22)$$

α_F and α_T are usually smaller than 1 and thus is the product and its squareroot are smaller than 1. The aspect ratio of the eddies at the onset of convection is greater than 1, i.e. their horizontal extension is greater than their vertical extension. Table 5.2 shows the characterization of the marginal state for some typical realizations.

The margin for the onset of convection obviously depends very much on the anisotropy which characterizes the system. If parameters differ by orders of

---

[2] Only in this subchapter α_T denotes *not* transversal dispersivity!

magnitude, the resulting critical Rayleigh numbers will do so as well. Note that for Ra^z, the classical value of $4\pi^2$ remains the marginal level, if anisotropy ratios for permeability and diffusivity are equal - independent of the ratio's value.

The eddies at the onset of convection become wider as anisotropy increases. The aspect ratio of 10 is reached if there is an anisotropy factor of 1/100 for both parameters considered.

α_F	α_T	Ra^*	Ra^x	Ra^z	ϑ	$1/\vartheta$
1.0	0.1	17.098	17.1	171.0	0.562	1.778
0.1	1.0	17.098	1.71	17.10	0.562	1.778
1.0	0.01	11.942	11.94	1194.	0.316	3.165
0.01	1.0	11.942	0.119	11.94	0.316	3.165
0.1	0.1	3.948	0.395	39.48	0.316	3.165
0.1	0.01	1.710	0.171	171.0	0.178	5.623
0.01	0.1	1.710	0.017	17.10	0.178	5.623
0.01	0.01	0.395	0.004	39.48	0.100	10.0

Table 5.2: Characterization of marginal state for anisotropic situations

Epherre (1975) treated the anisotropic case and obtained the same results as presented above. The general behavior of porous medium convection is similar for pure fluids. Anisotropy in eddy diffusivity and viscosity causes a change of the critical margin and a widening of the rolls in the supercritical situations. Further details may be found in the publication of Ray (1965).

5.3 Bifurcation Analysis

Bifurcation analysis enables the determination of branching of solutions in the phase space. The problem is transformed into a set of ordinary differential equations with linear part $\mathbf{A} \cdot \mathbf{u}$ and nonlinear part $\mathbf{f(u)}$:

$$\frac{\partial}{\partial t} \mathbf{u} = \mathbf{A} \cdot \mathbf{u} + \mathbf{f(u)} \qquad (5.23)$$

The behavior of the solution $\mathbf{u}$ near an equilibrium is characterized by the eigenvalues of the matrix $\mathbf{A}$. An eigenvalue λ is generally a complex number and the real part determines whether a perturbation is damped ($\text{Re}(\lambda)<0$) or amplified ($\text{Re}(\lambda)>0$).

In this subchapter I present again a stability analysis based on the streamfunction formulation for 2D thermal convection in porous media. It is shown that the well-known results on the stability of the different solutions can be obtained by the proposed approach. The system of ordinary equations is derived from the a simplified version of the perturbation equations (5.5):

$$\nabla^2\Psi^* = -Ra\frac{\partial\theta^*}{\partial x}$$

$$\nabla^2\theta^* + \frac{\partial\Psi^*}{\partial z}\frac{\partial\theta^*}{\partial x} - \frac{\partial\Psi^*}{\partial x}\frac{\partial\theta^*}{\partial t} = \frac{\partial\theta^*}{\partial t}$$

(5.24)

Note that one of the terms containing a factor of perturbations has been omitted, while the other has been kept. The following stability of the convection roll patterns is guided by an analysis made by Bhattacharjee (1987) for convection in pure fluids. The following formulas are assumed to be valid with unknown time-dependent amplitudes $a(t)$, $b(t)$ and $c(t)$:

$$\Psi^* = -\frac{a}{\pi}\sin(a_0 x)\sin(\pi z)$$

$$\theta^* = b\cos(a_0 x)\sin(\pi z) + c\sin(2\pi z)$$

(5.25a,b)

Concerning the last term of (5.25a,b) in comparison with (5.7) this is in fact an extended approach using the spectral formulation (5.9). But for the mode-numbers s in vertical direction special values are selected. In fact the additional term in (5.25a,b) is the lowest mode, which contributes to heat transfer. Introducing these formulae into the streamline equation above and comparing coefficients leads to the equation:

$$\left(a_0^2 + \pi^2\right)\frac{a}{\pi} + Ra\cdot ba_0 = 0$$

(5.26)

The transport equation is transformed in two ways. First it is multiplied by $\cos(a_0 x)\sin(\pi z)$ and then integrated over x over one cell (from $-\pi/a_0$ to π/a_0) and over z from 0 to 1. In order to obtain a second equation it is multiplied by $\sin(2\pi z)$ before integration. In literature on Finite Element or spectral methods this procedure is referred to as Galerkin method (Schwarz 1984, Canuto e.a. 1988). The following two equations result:

$$\dot{b} = -b\left(a_0^2 + \pi^2\right) - \frac{a}{\pi}a_0 - ac$$

$$\dot{c} = -4\pi^2 c + \frac{1}{2}ab$$

(5.27)

The dot on top of the letters on the left hand side of the equations denotes a time derivative. Defining $b_0 = \pm\sqrt{a_0^2 + \pi^2}$ the relationship $a = -Ra\pi a_0 b / b_0^2$ is used to simplify equations (5.27) to a system for two unknown function b and c. The timescale transformation $t \rightarrow b_0^2 t$ leads to:

$$\dot{b} = -(1 - R\frac{a_0^2}{b_0^4})b + Ra\frac{a_0^2\pi}{b_0^4}bc$$

$$\dot{c} = -\frac{4\pi^2}{b_0^2}c - Ra\frac{\pi a_0}{2b_0^4}b^2$$

(5.28)

With the definitions $r = Ra \cdot a_0^2 / b_0^4$ and $s = a_0 / \pi$ the following system results:

$$\dot{b} = (r-1)b + \frac{r}{s}bc$$

$$\dot{c} = -\frac{4}{1+s^2}c - \frac{1}{2}\frac{r}{s}b^2$$

(5.29)

This is the type of system that was to be obtained. With

$$\mathbf{A} = \begin{pmatrix} r-1 & 0 \\ 0 & -\frac{4}{1+s^2} \end{pmatrix} \quad \text{and} \quad \mathbf{f(u)} = \begin{pmatrix} \frac{r}{s}bc \\ -\frac{r}{2s}b^2 \end{pmatrix} \quad \text{for} \quad u = \begin{pmatrix} b \\ c \end{pmatrix} (5.30)$$

the form (5.23) is given. Steady solutions are obtained first. When time derivatives are set to zero, the first equation leads to:

(1) $b = 0$ or (2) $c = -(r-1)\frac{s}{r}$ (5.31)

Case (1) requires $c=0$ as well; the trivial conductive state is obtained. In case (2) two convective states exist, if there are two roots for b:

$$b = \pm\sqrt{\frac{8s^2}{1+s^2}\frac{r-1}{r^2}}$$

(5.32)

The condition for the existence of the roots is obviously $Ra > b_0^4 / a_0^2 = (a_0^2 + \pi^2)^2 / a_0^2$. In order to find the minimum value for Ra and the respective a_0 the minimum of the function $Ra(x) = (x + \pi^2)^2 / x$ has to be analyzed. The derivative $Ra'(x) = (1 - \pi^2 / x^2)(1 + \pi^2 / x^2)$ has one zero only: $x = \pi^2$. Thus the minimum value for Ra - the critical Rayleigh-number - is $4\pi^2$ (or $r=1$) and the wave number at the onset of convection is π (or $s=1$).

This characterization of the marginal state is the classical result of Lapwood (1948). Nevertheless, in contrast to the classical analysis based on a set of partial differential equations the classical statement is derived here from a system of ordinary differential equations.

Physically, the non-trivial solutions are the 2D-convection rolls with different rotation. The stability of the steady states can again be analyzed on the linearized system. Variations δb and δc in the vicinity of (b_0, c_0) fulfill the two equations:

$$\dot{\delta b} = (r-1)\delta b \qquad\qquad \dot{\delta c} = -\frac{4}{1+s^2}\delta c \qquad (5.33)$$

The trivial steady state is stable for $r<1$. For $r>1$ it becomes unstable: perturbations δb increase. The growth rates for the modes $a_0=\pi$ and $a_0=\pi/2$, as calculated in equation (5.33) differ from those given by Leijnse/Oostrom (1994) only by the factor, which stems from the transformations of the equations.

The same type of analysis is performed for the two convection states (b_1, c_1) and (b_2, c_2). The following system of linear equations appears:

$$\begin{pmatrix} \dot{\delta b} \\ \dot{\delta c} \end{pmatrix} = \begin{pmatrix} r-1+\dfrac{r}{s}c_{1,2} & \dfrac{r}{s}b_{1,2} \\[2mm] -\dfrac{r}{s}b_{1,2} & -\dfrac{4}{1+s^2} \end{pmatrix}\begin{pmatrix} \delta b \\ \delta c \end{pmatrix} \qquad (5.34)$$

The eigenvalues of the matrix are identical to the coefficients in the exponential functions $\exp(\mu t)$ and they are given by the formula:

$$\mu_{1,2} = -\frac{1}{1+s^2}(2\pm\sqrt{4-8(r-1)(1+s^2)}) \qquad (5.35)$$

For $s=1$, the wave number at the onset of convection follows: $\mu_{1,2} = -1\pm\sqrt{1-4(r-1)}$.

The region of interest, where $r>1$ holds, can be divided into three classes concerning the type of the eigenvalues:

$1<r<1.25$	two different real valued eigenvalues
$r=1.25$	double eigenvalue $\mu=-1$
$1.25<r$	eigenvalues with $Re(\mu)<0$ and $Im(\mu)\neq 0$

Both equilibria are stable for all $r>1$. For small supercritical values of r, the steady state solution is approached directly, if perturbations are small. For higher values the transient solutions spirals in to the stable point.

Altogether the preceding analysis shows that for values above the critical Rayleigh number there are three solutions. Two represent steady convection and are stable; the third solution representing conduction only, is unstable. Thus, when starting from the subcritical values Ra is increased and the critical margin is crossed, the there is an *exchange of stability* from conduction to convection solutions. Fig. 5.3 gives a graphical representation of that event. In bifurcation theory this is called a *pitchfork bifurcation*.

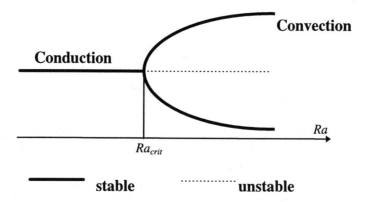

Fig. 5.6: Exchange of stability - illustration of the pitchfork bifurcation

The result has been obtained under the assumption that the perturbations have a certain form. In chapter 7 it will be shown that the steady convection solutions become unstable, if the Rayleigh number is increased further. This shows that the analysis of this chapter is valid only in the vicinity of the first critical Rayleigh number.

In this subchapter porous media convection has been studied based on an analysis of a set of ordinary differential equations. An analogous treatment of convection in pure fluids has led Lorenz (1963) to the discovery of deterministic chaos. The interested reader finds a good introduction by Bhattacharjee (1987). In the pure fluid case a set of three differential equations remains, while there are two equations in porous media convection (sets (5.27), (5.28) or (5.29)). That makes a fundamental difference: the route to chaos in porous media is different from pure fluid chaos. Some additional remarks are given in chapter 7.1.

5.4 Numerical Experiments

There are numerous publications of numerical calculations on free convection in porous medium. There are three main paths, which are followed by researchers in that field. In fact spectral methods (1) are the technique applied most, using truncated Fourier-series as basis and weighting functions (see: Galerkin approach, chapter 5.3). Beyond that simulations using standard Finite Difference (2) and Finite Element (3) methods can be found. The main early investigations are reviewed and described by Combarnous/Bories (1975), Caltagirone (1982) and Bories (1987). Basic equations, discretization methodologies and solution techniques are quite different and cannot be tackled in detail here. This chapter presents some simulations with the FAST-C(2D) code.

5.4.1 Isotropic Porous Medium

Convection Patterns

The 1st mode convection patterns in a 2D square cell was studied with a model set up using FAST-C(2D). An equidistant grid with 40 block in both directions was chosen. The accuracy of the numerical method was set to 10^{-4} - for linear solvers and Picard iteration. The discretization method for 1st order terms was central weighting.

Steady state patterns have been obtained by a transient simulation from an initial state representing a slightly disturbed conductive state. Timesteps were chosen in order to fulfill the Courant-criteria and the Neumann criterion (see chapter 4). Numerical steady state is reached when the state variables do not change in a Picard iteration. Some RESTART simulation runs with decreasing timesteps showed that the solutions change only marginally. The numerical steady state was taken as steady state.

Ra was selected to be 40, 80, 120, 160, 200, 240, 280, 320 and 360[3]. Note that for lower Rayleigh numbers no convection solutions exist. When *Ra* exceeds 380 the system becomes unstable (see chapter 7).

The streamline pattern reveals one circulating eddy in all simulations. The resulting isotherms within the cross-section from the simulations with different Rayleigh numbers are shown in Fig. 5.7. Each of the nine single figures shows contours of the dimensionless variable θ. Contour line levels are 0.05(0.05)0.95. For *Ra*=40 the picture is relatively simple: isotherms from the left hand to the right hand side move from higher depth to lower depth below the top boundary.

Note that the convection roll has a counter-clockwise rotation here. Rotation depends on the initial disturbance. Thus similar figures in which left and right side are exchanged are solutions as well. The mirrored figures can be obtained by changing the initial disturbance analogously.

At *Ra*=80 the isotherm picture becomes more complicated. Isotherms starting on the left side in their upward movement turn back in the vicinity of the model center before the final turn towards the right hand side. Sometimes this is referred to as *S-shape*. Some authors call this the build-up of *boundary layers*. Straus (1974) describes it as follows: '*this allows the interior of the fluid to approach isothermality, while the heat flux across the boundaries is increased owing to large temperature gradients there*'. This has the following effect on temperature observations along the vertical line through the center: starting from the top the temperature first increases, then decreases in a small intermediate part, and increase again towards the bottom.

Fig. 5.7 shows clearly that this effect can be observed only near the center of the model region. The intermediate interval, in which a decrease of temperatures can be observed, becomes smaller when the vertical observation line is moved

---

[3] The input data for an equivalent model with *Ra*=100 can be found on CD-ROM in the file *Benard.WTX*

from the center towards the boundaries, and vanishes at a certain distance. The described effect gains strength with increasing Rayleigh number. When $Ra=360$, all isotherms between $\theta=0.3$ and $\theta=0.7$ show a strong turn back.

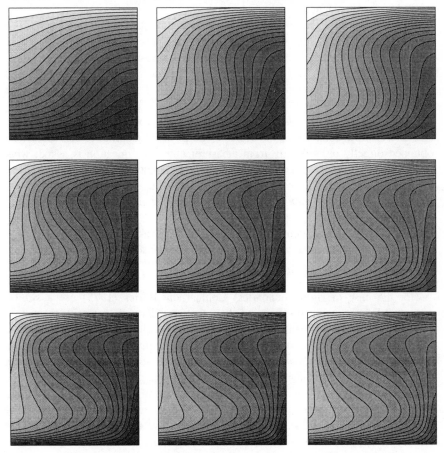

Fig. 5.7: Isotherms for classical Bénard-convection: results from numerical simulation with FAST-C(2D) with changing Rayleigh number
top row: $Ra=40, 80, 120$; center: $Ra=160, 200, 240$; bottom: $Ra=280, 320, 360$

Sensitivity to the Discretization Method

All three discretization methods have been applied for various Rayleigh numbers near the marginal state in order to check their accuracy by comparison with theoretical results from weak nonlinear analysis. The transient behavior of the first mode case with square convection cells was traced until steady state was reached. The computations used an equidistant grid of 32×16 blocks for a rectangular geometry with $H=1$ and $L=2$. This discretization ensured that the grid-Péclet number criteria (4.14) were fulfilled for all runs with $Ra<80$.

Geometrical conditions remained the same. Ra is the physical parameter which was changed. Results were checked by simulation runs started with different initial conditions. For a fixed value of Ra the two-eddy steady state was obtained by more than one transient simulation.

Heat transfer through the system is characterized by the Nusselt number Nu. For Nu definition (3.63b) can be used on the closed upper and lower boundaries. In the static case, where there is only conduction, Nu is equal to 1. In the supercritical case Nu is greater than 1. The Nusselt number thus describes the increase of total transfer through the system related to the transfer of the conductive state (see chapter 3).

Ra characterizes a system by its physical properties. In contrast Nu characterizes the behavior of the system. The physical parameters can usually be obtained directly or indirectly from measurements. The question is asked: what will be the mean transfer under the given physical conditions. The *inverse problem* has the response of the system as a basis and aims to determine one of the parameters. Both cases require an understanding of the connection between Nu and Ra numbers.

Concerning thermal convection in pure fluids Nakagawa (1960) derived a simple formula relating Ra and Nu numbers near the onset of convection:

$$Nu = 1 + 2 \left(Ra - Ra_{crit} \right) /Ra \qquad (5.36)$$

(for the case of rigid boundaries) Combarnous/Bories (1975) use a similar treatment for porous media flow and show that equation (5.36) holds there, too. The relation is independent of the wavenumber because it is gained under the assumption *'that the pattern of motions will be identical with that given by the solutions of the linear theory'*. As Nakagawa (1960) notes, the formula is true in the neighborhood of the marginal state and it is true only for the base mode.

Perturbations of Rayleigh numbers were used by Palm e.a. (1972) to show the Nu-Ra-relationship in 2nd, 4th and 6th order. The formula near the onset of convection is in fact the same as given by Nakagawa (1960) - as Cheng (1978) already noted. Zebib/Kassoy (1978) show that equation (5.36) can be generalized for single layer convection cells by replacing critical Ra by Ra_{crit}^{rl} .

Fig. 5.8 illustrates the output of Nusselt numbers (cubic approximation) calculated by the FAST-C(2D) model together with the approximations of 2nd, 4th and 6th order from Palm e.a (1972). BIS-simulations give the smallest values for Nu. This is due to increased diffusion/dispersion in the numerical system, i.e. Ra of the simulated system is smaller than the input Rayleigh number. The effect is reduced by using the truncation error correction. For low Ra the CIS scheme results match best with the theoretical curve. The coincidence is better if the monotonous increase of the approximations derived from theory (see Palm e.a. 1972) is taken into account. This is the reason why the CIS discretization is preferred in most calculations presented in this chapter.

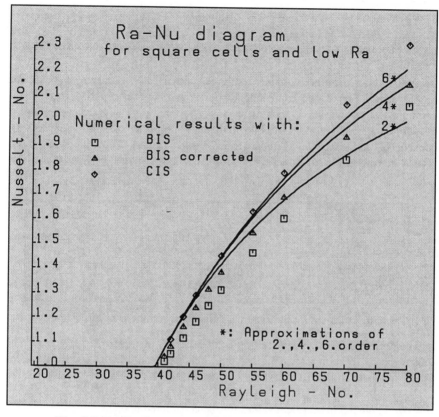

Fig. 5.8: *Ra-Nu* diagram for square cells near the marginal state

Ra-Nu-Diagram for Different Modes

A series of numerical simulations of the transient development of the nonlinear is carried out in order to show the different branches of steady state solutions - both in the vicinity and distant from the marginal state. Five branches of one-layered (2D-) eddy solutions for different cell aspect ratios ϑ are examined. The height-to-length ratios ϑ for the steady state cells are 1., 1.5, 2., 2.5 and 3.

The different branches of the solution (with different cell aspect ratios ϑ) are traced in the following way: the steady state solution for a special value of ϑ is used as an initial value for a simulation run with a slightly differing value for *Ra*. The new steady state is again taken as initial state for another *Ra*. This procedure makes it possible to follow some branches up to *Ra*=500 and down to *Nu*<2. The results are plotted in Fig. 5.9. For high Rayleigh numbers there is a transition to an oscillating regime (see chapter 7). If Nusselt numbers approach 1, the changes of state variables are very small, even for large timesteps. It becomes impossible, to distinguish between steady state and slow transient state.

For some values of Ra and ϑ the obtained Nusselt numbers can be compared with experimental and numerical findings. The outcome corresponds well with results published by Caltagirone (1975) and Straus/Schubert (1979b).

All branches of the solution - characterized by the cell aspect ratio - show an increasing heat transfer with increasing Ra, described by curves of parabolic shapes. A solution traced from high Rayleigh numbers downward, changes at a certain point to another branch with smaller r and ϑ. Therefore the branches cannot be traced down to $Nu=1$. The numerically observed instability confirms the concept of *secondary bifurcation* (Riley/Winters 1989). Only the square cell branch can be followed down to the marginal value $Nu=1$ - there is no other solution to which the system could switch.

No. of cells r	2	3	4	5	6
Aspect ratio ϑ	1.	1.5	2.	2.5	3.
Ra_{crit}^{rl}	39.5	46.	62.	83.	110.

Table 5.3: Critical Ra for different cell aspect ratios (from linear theory)

The toe point of the square cell branch (i.e. $\vartheta=1$) is at $(Ra=Ra_{crit}, Nu=1)$. The (not reached) toe points of the other branches are identical to the critical values Ra_{crit}^{rs} from equation (5.13), where r is the number of cells in horizontal direction in the system with $H=1$ and $L=2$. This can be seen easily by using the relationship: $\vartheta = r/L$. Table 5.3 lists critical values for the considered branches.

Fig. 5.9 illustrates the critical margins with vertical broken lines. The branches are headed towards these points, although they do not reach them - with the exception of the square cell branch. In systems which are slightly supercritical heat transfer is higher for smaller aspect ratios ϑ and wider rolls. With growing Ra the situation becomes different: Nu then increases with ϑ and r, i.e. slender rolls favor higher heat transfer. While for low Ra the square cell branch is the mode with highest Nu, for higher values of Ra it becomes the one with the lowest Nu.

For example: at $Ra \approx 165$ the heat transfer of the 1st and 2nd mode ($\vartheta=1$ and 2) become equal. For higher Ra the Nusselt number of the 2nd mode exceeds that of the 1rst mode. Note that the Rayleigh numbers of these 'crossing points' are not identical with Ra for modal exchange (for the 1st and 2nd mode exchange occurs at $Ra = 9\pi^2/4 \approx 45$,see Riley/Winters 1989).

Most publications show Ra-Nu-diagrams with single curves and do not mention that the relation between the two dimensionless numbers is only valid for a certain flow pattern. These simple graphs are shown even though it is well understood that the Ra-Nu-relation is not unique: it depends on the cell pattern and it should be clearly indicated which cell formation is considered. The Ra-Nu-diagrams in this book mostly show an ensemble of curves, each one representing one convective mode.

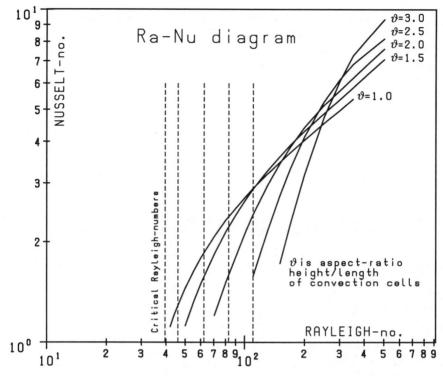

Fig. 5.9: *Ra-Nu* diagram for five 2D-cell solution branches

The observation of the numerical solutions shows that once the system has reached a certain branch, it stays on that branch, even if some parameters change to values very different from the starting point. So the system does not necessarily switch to the cell pattern belonging to the lowest possible Rayleigh number!

Modal exchange is a term to distinguish regions with different convective flows at the onset of convection. These points are not of practical interest, as the Rayleigh number is often much higher than in the marginal state. Nevertheless, if *Ra* approaches $4\pi^2$ from above, all patterns except the critical mode pattern become unstable. Then modal exchange can be observed. Riley/Winters (1989) therefore present corresponding modal exchange points and Rayleigh numbers together.

Platzman (1965) stated that the solution with maximum heat transfer is preferred. The Nusselt numbers in real systems would be expected slightly above those curves shown in Fig. 5.9. Numerical systems have not been observed to have such a preference. As mentioned above, several solutions exist for the same situation. Even if some physical parameter changes, the system does usually remain in the same mode. Moreover, as noted above, Rayleigh numbers characterizing modal exchange do not coincide with Rayleigh numbers where heat transfer of two modes is equal.

At this point a remark is to be made on 3D-effects, which have been studied in a few publications. Holst/Aziz (1972) note that 3-dimensional convection *'exhibits considerably more heat transfer'* for high *Ra*. Low *Ra* systems have the reverse situation (Zebib/Kassoy 1978). The curves for 3D-convection branches may be even steeper than the steepest shown in Fig. 5.9.

Sensitivity on Grid Spacing

A sensitivity study was conducted concerning the spatial grid spacing. Three runs were made using FAST-C(2D) code with equidistant grids of 20×20, 40×40 and 80×80 blocks. All physical or geometrical parameters were kept constant: $H=1.$, $L=1.$, $Ra=100$; timesteps as well. Table 5.4 lists and compares the main output parameters of the runs.

Characteristic	20 x 20	40 x 40	80 x 80
θ_{min}	0.0116	0.0567	0.0278
θ_{max}	0.0988	0.9943	0.9970
Nu, 1.order, top	2.4327	2.5317	2.5918
Nu, 1.order, bottom	2.4328	2.5318	2.5916
Nu, 2.order, top	2.4746	2.5432	2.5948
Nu, 2.order, bottom	2.4747	2.5434	2.5946
Nu, 3.order, top	2.4513	2.5351	2.5926
Nu, 3.order, bottom	2.4513	2.5353	2.5924
Ψ_{max}	-0.0349	-0.0088	-.0022
Ψ_{min}	-5.3495	-5.3150	-5.3802
v_{max}	4.3508	4.3852	4.4201
$v_{x,max}$	18.842	19.069	19.458
$v_{x,min}$	-18.842	-19.068	-19.464
$v_{z,max}$	18.928	19.228	19.538
$v_{z,min}$	-18.930	-19.230	-19.538
$Pe_{x,max}$	0.9421	0.4767	0.2433
$Pe_{z,max}$	0.9465	0.4807	0.2442
$Cou_{x,max}$	0.038	0.078	0.159
$Cou_{z,max}$	0.038	0.081	0.168
Neu	0.08	0.32	1.28

Table 5.4: Characteristic values for convection at $Ra=100$ obtained with different grids

Minimum and maximum values of the transport variable naturally change with the grid. For finer grids minima must be smaller, maxima be higher, because the corresponding block centers lie nearer to the boundaries. For similar reasons maximum and minimum for streamfunction Ψ and velocity components show the same type of behavior. Grid-Péclet numbers Pe_x and Pe_z, Courant numbers Cou_x and Cou_z and the Neumann number all depend directly on the grid spacing (see chapter 4) and thus necessarily change with the grid refinement. The Nusselt number is the best parameter to give an estimate of the convergence of the numerical method. Nusselt numbers are defined integrally (see equation (3.63b)) and are extrapolated from values in the interior towards the boundary (see chapter 4.7).

For the mentioned runs there is still a change of 2.4% between the two finest grids. The difference is smaller than between the two coarsest grids, which is around 4%. The results given are valid within the accuracy of a few percent. Further grid refinements could reduce the observed uncertainty. Note that the differences for the different extrapolation orders reduce significantly with finer spacing. Note additionally that errors between top and bottom approximations are marginal.

Solver accuracy may also influence results. As mentioned above the accuracy dependent numerical steady state is taken as criterion for the physical steady state. Change of solver accuracy may cause a change in the results of some percent.

5.4.2 Anisotropic Porous Medium

The case of anisotropy in permeability was studied numerically using the FAST-C(2D) code. Results are given in Table 5.5. Vertical permeability was 10 times smaller than horizontal permeability.

Table 5.5 presents the outcome of a sensitivity-analysis for the numerical model due to changes of the Rayleigh number in the vicinity of the marginal state. The Nusselt number and the maximum value of the streamfunction Ψ_{max} of the final state are shown in the second and third column. The fourth column notes the time, at which the end criterion of the simulation was reached. The last two columns contain maximum grid-Péclet numbers in both coordinate directions in the final state.

Ra	Nu	Ψ_{max}	t_{max}	Pe_x	Pe_z
23	1.398	2.979	1.8	0.605	0.335
22	1.330	2.654	1.8	0.535	0.298
21	1.260	2.300	2.1	0.460	0.257
20	1.187	1.905	2.7	0.379	0.213
19	1.115	1.446	4.5	0.285	0.161
18	1.042	0.850	7.5	0.166	0.095
17	1.000	0.063	.25	0.012	0.007

Table 5.5: Results from modeling the case with anisotropy in permeability

The qualitative change of the regime is clearly demonstrated when *Ra* switches from 18 to 17. Very short simulation time belong to situations, where the Nusselt number is extremely close to 1.0. Streamfunction values and grid-Péclet numbers are significantly reduced compared to the convective states, which have been reached for higher Rayleigh numbers. This coincides perfectly with the marginal state which was determined by theoretical analysis as *Ra*=17.1 (see Table 5.2).

Another sensitivity analysis with a greater overall variability of *Ra* was made in a model of a rectangular porous medium box with a L/H-ratio of 4. The discrete grid consisted of 80×20 equally spaced blocks. The temporal simulation proceeded using the Crank-Nicolson method (see chapter 4). Timesteps were chosen automatically considering both Courant- and Neumann-criterion (with timestepping factor χ=1.0). The solver for flow was preconditioned CG with a relaxation factor of 1.7; transport was computed with preconditioned CGS and the same relaxation factor. The relative accuracy required in each iteration was usually ε=10^{-5}. Only in few cases it was difficult to reach a steady state with this accuracy. In that case values up to ε=10^{-4} were used for the iterations in the numerical model.

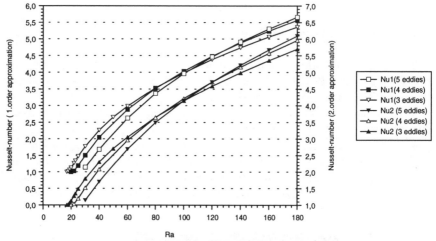

Fig. 5.10: Nusselt numbers for anisotropic porous medium (k_x/k_z=10) - calculated by 1st and 2nd order approximations from numerical output of FAST-C(2D)

Fig. 5.10 and Fig. 5.11 show *Ra-Nu*-diagrams for the anisotropic porous medium case. As mentioned above, the quantity of increase of *Nu* with *Ra* depends on the solution branch, i.e. the type of convection cells. For *Ra*=10 the cases with 3, 4 and 5 eddies respectively in horizontal direction were considered. The corresponding cell aspect ratios are ϑ=0.75, 1.0 and 1.2 respectively. For *Ra*=100 cellular patterns with 2, 3 and 4 eddies were studied (ϑ=0.5, 0.75 and 1.0). The branches were traced starting with *Ra*=180 or *Ra*=120 downward.

Nusselt numbers can be evaluated with different approximations, as shown in chapter 4.7. Here 1st and 2nd order formulae are used - abbreviated as Nu1 and

Nu2. Note that in both figures the origin of the ordinate is shifted by 1.0 depending on the order of approximation. Otherwise it would be difficult to distinguish the six curves.

Fig. 5.10 and Fig. 5.11 show ensembles of *Ra-Nu* curves. The critical value characterizing the marginal state, can be taken from the position of the toe point nearest to the origin of the coordinate system. Here it is the toe point of the 1st mode branch (2 eddies) in both figures. The result is, that for the low anisotropy case the critical Rayleigh number is close to 17; in the high anisotropy case it is close to 12. According to Table 5.2 the critical values are 17.1 and 11.94, which has been analytically derived. Numerical and analytical output fit together perfectly.

The analyses show the limits of the numerical method. Different approximations used to calculate *Nu* (mean heat or mass transfer through the system) lead to differences up to 0.43 when the anisotropy ratio is 10 and 0.65 when the ratio is 100. This uncertainty is obviously in the same range as the differences between the branches. Approximations of higher order lie in between the margins set by the 1st and 2nd order approximation. It seems reasonable to use a mean of both or the 3rd order approximation as a better estimation of *Nu*.

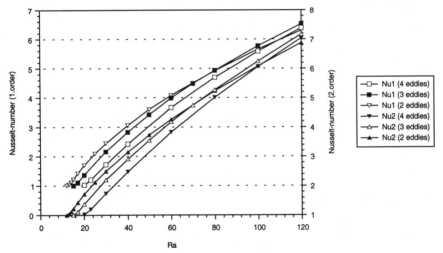

Fig. 5.11: Nusselt numbers for anisotropic porous medium (k_x/k_z=100) - calculated by 1st and 2nd order approximations from numerical output of FAST-C(2D)

Convection in an anisotropic porous medium can be recommended as a benchmark case for density driven flow modeling codes.

6 Special Topics in Convection

6.1 Thermal Convection in Slender Boxes

6.1.1 Analytical Studies

Chapter 5 shows how convection rolls can be derived using the following approach for the variable functions(compare equations (5.7)):

$$\sin(r\pi x / L)\sin(s\pi z / H) \quad \text{and} \quad \cos(r\pi x / L)\sin(s\pi z / H) \tag{6.1}$$

r denotes the number of rolls in horizontal direction and s the number of cell layers in vertical direction. The total number of rolls is $n \cdot s$ and all eddies are of equal size. The aspect ratio of cell height to cell length is rH/sL (see eq. (5.8)).

It has been shown that a critical Rayleigh number can be introduced for each roll pattern; the formula was given in equation (5.12). The lowest critical value is given for $s=1=r/L$. If $\vartheta=r/sL$ denotes the cell aspect ratio in the unit system, the pattern-specific Rayleigh number can be denoted by:

$$Ra_{crit}^{s,\vartheta} = (1+\vartheta^2)^2 \pi^2 s^2 / \vartheta^2 \tag{6.2}$$

The well-known critical margin of $4\pi^2$ results from (6.2) setting $s=\vartheta=1$. These patterns can be realized in systems where the length L is a multiple of the height H. Most authors are interested in systems, whose horizontal extension is much longer than the vertical - the original experiments of Bénard (1900) had an aspect ratio $L/H=100$. The lowest value of the critical Rayleigh numbers for these systems is very near above the $4\pi^2$. The illustrative figure published by Beck (1972) shows that the critical Rayleigh number is slightly increased, if $L>H$, but not a multiple of H.

It is a totally different situation if $L<H$. Cells of quadratic shape extending over the whole vertical height are not possible. The critical Rayleigh number is higher than $4\pi^2$.

The following considers two different solution types. Type (A) consists of a long slender roll (one cell). Type (B) consists of a system of cells (multi cell). This first analysis assumes that the height is a multiple of length: $H=mL$. The critical Rayleigh numbers for both systems can be obtained with help of formula (6.2):

$$(A) \quad Ra_{crit}^{m,1} = 4\pi^2 m^2 \qquad (B) \quad Ra_{crit}^{1,m} = \frac{(1+m^2)^2}{m^2} \qquad (6.3)$$

The first two rows in Table 6.1 show the critical values for solutions of type (A). The last two rows show critical values for type (B):

m	1	2	3	4	5	10	15	20	30
$Ra_{crit}^{m,1} / \pi^2$	4	16	36	64	100	400	900	1600	3600
$Ra_{crit}^{m,1}$	39.5	158	355	632	987	3948	8883	15791	35531
$Ra_{crit}^{1,m} / \pi^2$	4	6.25	11.1	18.1	27.0	102	227	402	902
$Ra_{crit}^{1,m}$	39.5	61.5	109	179	267	1007	2240	3968	8902

Table 6.1: Critical Rayleigh numbers for one-cell and multi-cell-solutions in slender geometry for different aspect-ratios m

A comparison of the rows in Table 6.1 makes clear, that the 1-eddy solution always has a smaller Rayleigh number. The following will prove that the 1-eddy solution has the lowest critical value for all 2D-solutions in a given system where $L<H$. This result is independent of the box aspect ratio.

The proof proceeds in two steps. First, solutions with the same number of layers are considered and it is shown that the Rayleigh number becomes a minimal, if there is only one roll in horizontal direction. The use of the above notations and the additional use of k for the number of eddies in horizontal direction give the cell aspect ratio as $\vartheta = m \cdot k / s$. The marginal state derived with (6.3) is characterized by:

$$Ra_{crit}^{s,\vartheta} = \left(1 + \frac{m^2 k^2}{s^2}\right)^2 \frac{s^4}{m^2 k^2} \pi^2 \qquad (6.4)$$

As s is fixed, this expression is considered as a function of $w = k^2$. The derivative of (6.4) shows that the function increases monotonically in the interval $w \in [1,\infty[$. The minimum is obtained for $w=1=k$. This concludes the first part of the proof.

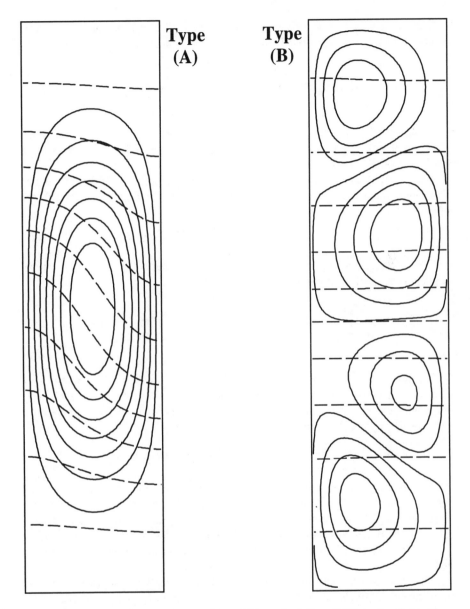

Fig. 6.1: Convection patterns of type (A) on the left and (B) on the right side

The broken lines are isotherms. Unbroken lines are streamfunction contours. Contour levels are selected equidistant. Both patterns are drawn from FAST-C(2D) outputs of transient simulations in a system with $H/L=4$ with different initial states. Later times of the same runs show that the one-cell flow pattern (A) remains, while the multi-cell system (B) vanishes.

The second step restricts the analysis to patterns where various numbers of rolls lie above each other. The critical value for these systems is given by:

$$Ra_{crit}^{s,\vartheta} = \left(1+\frac{m^2}{s^2}\right)^2 \frac{s^4}{m^2}\pi^2 = \left[m^2+s^2(2+\frac{s^2}{m^2})\right]\pi^2 \qquad (6.5)$$

Equation (6.5) is recognized Here as a function of s^2. It is a quadratic polynomial with a minimum for a negative argument. The minimum for all positive integers is obtained for $s=1$.

Beck (1972) considers 2D and 3D convection patterns. The functions which describe cellular systems in a box with length L_x, width L_y and height 1.0, are given by:

$$\sin(s\pi z)\cos(n\pi x/L_x)\cos(l\pi y/L_y) \qquad (6.6)$$

2D-rolls are obtained from this approach by choosing $l=0$ or $n=0$. The critical Rayleigh number according to Beck is given by:

$$Ra_{crit}^{s,n,m} = \frac{\pi^2}{(n^2+l^2)^2}\left(s^2+\frac{n^2}{L_x^2}+\frac{l^2}{L_y^2}\right)\left(L_x^2 n^2 s^2+L_y^2 l^2 s^2+(n^2+l^2)\right)$$

$$(6.7)$$

Formula (6.3) results when $l=0$. Formula (6.4) results, when $n=0$. Considering convection, the equation needs to be studied for cells which have horizontal extensions much smaller than the vertical extension. When horizontal grid spacing approaches zero, holds:

$$\lim_{L_x,L_y\to 0} Ra_{crit}^{s,n,l} = \infty \qquad \text{for fixed values } s,n,l$$

This is valid for all convective systems as well as for the minimum of the Rayleigh numbers. The minimum of specific Ra's for all 2D- and 3D-patterns is given for the 2D cases. The one cell solution will appear near the critical margin. The axis of rotation is parallel to the shortest of both horizontal spacings $L_x<1$, $L_y<1$. 2D-rolls are the ones which destabilize the conductive system for lowest values of Ra.

This is in fact the analogue to Squire's theorem (1934). It states for the pure fluid case that '*for the study of the stability between parallel walls it is sufficient to confine attention to disturbances of two-dimensional type*'. One conclusion from this result is: in order to study the onset of convection numerically, it is sufficient to study 2D convection rolls. These are the solutions which destabilize the system towards convection at lowest Rayleigh numbers.

6.1.2 Numerical Experiments

Numerical experiments are carried out applying FAST-C(2D) code. The aim is to compare analytical and numerical results. Simulations of the transient development of convective patterns are performed as in chapter 5.4.

The initial conditions chosen is the hydrostatic situation with constant streamfunction ($\Psi\equiv0$). The initial temperature is layered with colder water at the top and warmer water at the bottom. Some singular perturbations are added to start the simulation. Those perturbations are selected with the intention of favoring certain cell patterns. As an example: two deviations from the layered temperature distribution at opposite edges of the box promote the emergence of a 2-eddy system in the first part of the simulation. If the perturbations have opposite sign (one is higher - the other lower than the local environment), the two eddies will have opposite circulation. If the perturbations have equal direction, the appearing rolls will show the same rotation.

The patterns shown in Fig. 6.1 were produced by simulations with FAST-C(2D). The single eddy flow appears from a single perturbation at medium depth near the left boundary. The four eddy system was initiated by four perturbations - two near the left boundary and two near the right boundary. Two perturbations represent hot spots, two other represent cold spots in the local vicinity.

The temporal development of these patterns is of interest. There are three possibilities investigated:

(1) the system moves towards the conductive state (the perturbation dies out)

(2) the favored flow pattern remains (stable supercritical case)

(3) the pattern favored by initial conditions switches into a different convective state after a certain period of time and then remains stable

Numerical simulations give information on the stability of flow patterns. Numerical simulations also indicate, if the system is sub- or supercritical.

All simulations reported here ran into a stable steady-state. As criterion for the steady-state it was required that the numerical solution does not change within the last timestep. With this criterion the simulation reaches a steady state and cannot not be continued unless numerical parameters are changed. Those parameters are the timestep and accuracy criterion for the Picard iteration (compare chapter 4.5.2). Automatic time-stepping was chosen for the runs reported here. The timestep was determined from the Courant-criterion (with timestepping factor $\chi=.25$ - compare chapter 4.2). The accuracy to end the inner iteration was $\varepsilon=10^{-4}$.

Flow patterns as shown in output figures cannot be used to decide, whether the system is sub- or supercritical. This has already been reported by Holzbecher (1991a). Cellular flow patterns remain visible in graphical representations of numerical results, even if the system develops towards the subcritical conductive state. There is not much difference between maximum and minimum values of the streamfunction.

Near marginal state isotherms or isohalines are almost horizontal. Nusselt numbers can be used to quantify the almost non visible deviations from the linear regime representing the subcritical state. *Nu* as defined by equation (5.22) becomes 1.0 for pure conduction. Heat transfer from the bottom to the top of the closed box is then occurs by diffusion only. If *Nu* is above 1.0, the system is supercritical. There remain cases for which the nature of the system can hardly be determined, when the heat transfer is less than 0.3% above the minimum value - deviations of that size can be observed in subcritical cases far below the critical margin.

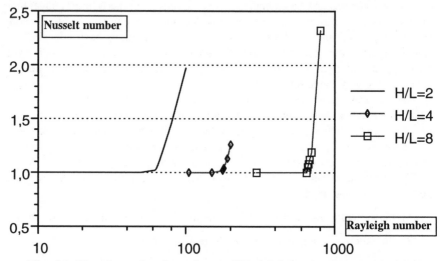

Fig. 6.2: Nusselt number dependency of Rayleigh number (supercritical cases are characterized by a Nusselt number greater than 1.0)

Under steady state conditions values for *Nu* at the upper and lower boundary are equal. This is also true for simulations made with the FAST-C(2D) code. Differences in heat transfer on both edges can be attributed to the initial conditions and thus decline gradually. The Nusselt numbers reported in the following were always determined by the third order approximation given in formulae (4.37) and (4.38).

Fig. 6.2 shows a *Ra-Nu*-diagram representing several numerical simulations with FAST-C(2D). Three systems have been modeled with different geometrical *H/L* aspect ratios. First a box with *H=2L* was modeled using a 20×20 grid with equally spaced blocks. In further model runs the aspect ratio *H/L* was changed to values of 4 and 8. Grids were refined to 20×40, 40×80 and 80×160 blocks[1]. As input parameter, the Rayleigh number was changed several times in all models in

---

[1] The input data for the model can be found on CD-ROM in the file *SlendCon.WTX*

order to trace the solution branch and find the margin for the onset of convection. Nusselt numbers were computed in a post-processing module using formula (6.17).

Fig. 6.2 shows clearly that from numerical runs marginal states can be determined within narrow bounds. The critical margins are very close to the values, which can be obtained analytically.

Critical values for the three models are 61.5, 178 and 652 as determined from equation (6.5). The correspondence between analytical and numerical results is a verification of the numerical code and a confirmation of the analytical derivation. The reader should remember that the analysis is based on the linearized system; numerical simulations are based on the nonlinear system. In spite of the method differences analytical and numerical results coincidence extremely well.

All simulations ended in single roll flow patterns - only the rotation direction is determined by the initial condition. The gradient of the isotherms or isohalines depends on *Ra*. This is illustrated in Figures 6.3 A to 6.3 F, where numerical results are shown for the same geometrical constellation, but for different Rayleigh numbers. The temperature gradient near the upper and lower boundary grows with increasing *Ra*. This gradient determines the total heat transfer through the system.

Figures 6.3 E and 6.3 F look almost identical, but there are indeed changes in the *Nu* number. For highly supercritical cases the temperature function is not monotonous on vertical lines near the middle axis of the system: following the line from the bottom, there is a drastic increase of temperature very near the boundary, while at a certain point temperatures start to decrease again. After another increase there is a final temperature drop when the upper boundary is approached. These S-shaped curves are already mentioned in chapter 5. In his study on saline convection Rubin (1975) refers to this effect as *'boundary layer formation'*.

Different initial states have been selected to check if multi-cell patterns vanish in all cases. This was done as described above. Initial perturbations favoring four cells, as shown in figure 6.1 B, died out developing to a single roll in the supercritical case. Variations with two layers of two rolls have also been checked, with the same result.

It can be concluded that computer runs nicely reproduce the behavior of the system, as it is known from mathematical analysis. Marginal points can be determined with a view on the *Ra-Nu* diagram. Unstable flow patterns at a certain point switch to the stable solution branch. Nonlinear system solutions behavior is quite well described by the analysis of the linearized system.

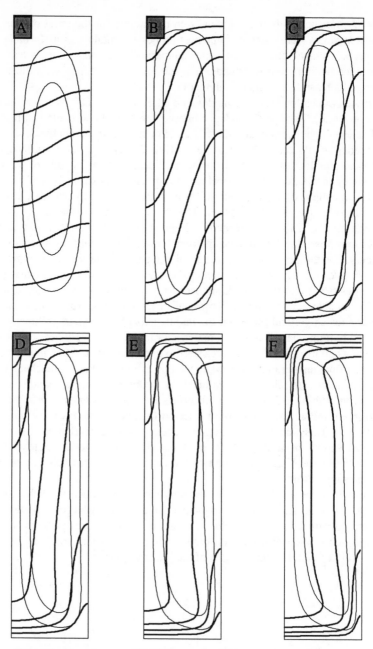

Fig. 6.3: Steady state 2D-flow patterns for a cavity with *H/L*=4
Ra equals: A, 200; B, 300; C 400; D 500; E 750; F 1000
Thin lines represent streamlines; thick lines represent isotherms; contours both for
streamfunction and temperature are for equidistant levels

6.2 Variable Viscosity Effects on Convection

6.2.1 Introduction

A study of the influence of viscosity in density driven flow is facilitated by some assumptions made to transform equation (3.39) into a more convenient form. An extended Oberbeck-Boussinesq assumption led to the formulation given in chapter 3. This chapter will study the isotropic and homogeneous case for thermal convection. If, additionally, density changes are assumed to be linear, then holds:

$$\frac{\partial^2 \Psi}{\partial x^2} + \frac{\partial^2 \Psi}{\partial z^2} = -Ra \frac{\mu_{ref}}{\mu(\theta)} \frac{\partial \theta}{\partial x} \qquad (6.8)$$

with
$$Ra = \frac{\gamma k \Delta \rho g H}{D \mu_{ref}} \qquad (6.9)$$

When μ is constant, the ratio of parameters in equation (6.9) - under the noted conditions - becomes a constant as well, which is the Rayleigh number. The subject of this chapter is non-constant viscosity. Ra is defined for a fixed value of reference viscosity. Equation (6.8) can be interpreted as follows: Ra is corrected locally by a correction factor $f_\mu = \mu_{ref}/\mu$ derived from the change of viscosity in relation to the reference value. The variable density effect, which is the origin of convective motions, is locally reduced, if the correction factor <1; or intensified, if the correction factor >1.

Note that a linear state equation $\rho(\theta)$ is assumed in the above formulation. The deviation from the linear dependency is very moderate for temperatures above 4°C - compared to the changes of viscosity. Temperatures near the freezing point require the density anomaly of water to be taken into account.

The dependency of viscosity on temperature has already been tackled in chapter 2.3.1. Only some additional remarks will be given here. The steam-tables of the ASME (Am. Soc. Mech. Eng.) according to Straus/Schubert (1977) and the JSME (Jap. Soc. Mech. Eng.) (1968) give the same approximating formula for the dependence of viscosity μ on temperature T (compare equation (2.4)):

$$\mu(T) = 241.4 \cdot 10^{247.8/(T-140)} \qquad (6.10)$$

The unit of dynamic viscosity is µPoise and T has to be given in °K. The formula is valid for fresh water in the range between 0°C and 300°C (ASME or JSME) and when p is the saturation pressure corresponding to temperature T. An extension of (6.10) for general pressure can be found in Straus/Schubert (1977). A formula considering variations of T and salinity c is used by SWIFT (1982):

$$\mu(T,c) = \mu_{ref} \exp \{a(c) (1/T - 1/T_{ref})\} \qquad (6.11)$$

Reference viscosity μ_{ref}, reference temperature T_{ref} and the function $a(c)$ have to be determined by measurements in the fluid. A similar functional approach is applied in the FAST-C(2D) code[2], where the dependence of μ on normalized temperature θ is implemented as follows:

$$\mu(\theta) = \mu_{ref} \exp \{ a [1/(\theta + \theta_0) - 1/(\theta_{ref} + \theta_0)] \} \qquad (6.12)$$

θ_{ref} and θ_0 are determined from two values $\mu_1 = \mu(\theta_1)$ and $\mu_2 = \mu(\theta_2)$ on the μ-curve. μ_{ref} is the reference viscosity used for the calculation of Ra. The slope parameter a is determined by the curvature of the viscosity graph. By increasing the constant a, the decrease of μ in the lower temperature part becomes steeper. Formulas for the calculations of θ_{ref} and θ_0 are

$$\theta_0 = - \frac{\theta_1 + \theta_2}{2} + \sqrt{[\frac{\theta_1 + \theta_2}{2}]^2 + \frac{a (\theta_1 - \theta_2)}{\ln(\mu(\theta_2) / \mu(\theta_1))}} \qquad (6.13)$$

$$\theta_{ref} = \theta_0 + 1 / (\frac{1}{\theta_1 - \theta_0} + \frac{\ln(\mu_{ref} / \mu_1)}{a}) \qquad (6.14)$$

As an example let us take the steam-table curve (6.10) in the temperature range from 20°C to 250°C. If both limits of the temperature interval are taken as θ_1 and θ_2, then $\theta_0 = -0.652$ holds. The slope parameter a can be calculated to be $247.8 \cdot \ln(10)/(250-20) = 2.48$. a depends on the temperature interval. With these choices, curve (6.12) fits exactly with equation (6.10). In fact, with convenient parameters (6.10), (6.11) and (6.12) are just three different formulas describing the same physical temperature dependence.

The general procedure, how functional dependencies of parameters are transformed in a dimensionless formulation, has been treated by Pawlowski (1991). He shows, that in the transformed systems only a certain class of functions χ has the characteristic, to be independent of the reference point (in the case of viscosity: independent of the reference temperature). This set of χ-functions can be noted in dependence of a parameter m:

$$\chi(u,m) = \begin{cases} (1 + m \cdot u)^{1/m} & \text{for } m \neq 0 \\ \exp(u) & \text{for } m = 0 \end{cases} \qquad (6.15)$$

where u is a transformed temperature. In fact (6.10), (6.11) and (6.12) are χ-functions with $m=0$, and equivalent to *Arrhenius*-functions. For temperatures

---

[2] The FAST-C(2D) version, included on CD, uses the same functional relationship to describe the dependence of viscosity μ on (not normalized) temperature T. Input values T_1 and T_2 are taken to calculate $\mu(T_1)$ and $\mu(T_2)$ in formulae (6.13) and (6.14).

between 0°C and 100°C equation (2.5) provides a better approximation with $m=-0.636$ (Pawlowski 1991).

Viscosity dependence on temperature is analyzed and modeled in some examples in the following. The concentration dependent case is not treated further. The influence of changing viscosity on saline convection can be modeled using FAST-C(2D).

Another related subject is the convection of non-Newtonian fluids. Amari e.a. (1994) use an approach very similar to the porous medium case to model 2D-boxes heated from below. In that case, viscosity becomes velocity-dependent and thus can be represented as a function of Ψ:

$$\mu = \left[\left(\frac{\partial \Psi}{\partial x}\right)^2 + \left(\frac{\partial \Psi}{\partial z}\right)^2\right]^{\frac{n-1}{2}} \qquad \text{with power - law index } n \quad (6.16)$$

FAST-C(2D) would require some minor changes in order to enable modeling convection of non-Newtonian fluids in that way. These have not been implemented yet.

6.2.2 Onset of Convection

The computer code FAST-C(2D) was used on a NEC-H9870 computer to test the onset of thermal convection. The size of the first timesteps was predefined; following timesteps were calculated automatically to fulfill the Courant-criterion (with timestepping factor $\chi=1.0$, compare chapter 4.2). The initial state of the simulation is a disturbed hydrostatic situation: the initial temperature increases with depth except for a disturbance in a single block. The simulation was stopped, when Ψ and θ did not change within a timestep simulation. The final state then is regarded as steady state.

An equidistant grid with 20×20 blocks was selected for a model with equal length and height. Some jobs have been run with a 40×40 block model in order to check the independence of the results from the grid-spacing. The grid-Péclet number criterion was fulfilled near the onset of convection even for the coarse grid. A check of the independence from the 1st order discretization scheme (backward) was made by some runs using the central scheme (compare chapter 4.1). Tests were also made to determine whether timestep size has influence on the code output.

The criterion, whether there is convection or not, was that Nusselt numbers along the vertical edges increase significantly from 1.0. The Nusselt numbers were calculated according to the following formula (compare equation (4.38)):

$$Nu = \frac{2}{L}\sum \frac{1}{\Delta z}(\theta - \theta_{bc}) \qquad (6.17)$$

where θ denotes the numerically calculated value for normalized temperature at the block-center, and θ_{bc} the value on the block-edge according to the boundary condition. The sum has to be taken across all boundary block edges on horizontal edges.

The sum has to be extended over all blocks at the lower or at the upper edge of the model. Theoretically Nu at steady state should be constant within each layer. Particularly the values obtained for upper and lower boundary should be equal. This is not true for calculated solutions with variable viscosity, while it was perfectly true for solutions with constant viscosity reported in the preceding chapters. In subcritical cases of varying viscosity, Nu at the upper model-edge was slightly above 1; at the lower edge it was slightly lower. Both cases deviate less than 0.1% from the hydrostatic value. In supercritical cases, both values were higher than 1. This characteristic behavior could serve as criterion for the onset of convection as well.

Table 6.2 shows Nusselt numbers for a constant viscosity case and a variable viscosity case. In the constant case μ was chosen 10 mPoise; in the variable case μ_{ref} was set to 2 mPoise. Low and high boundary temperatures were set to 20°C and 268°C, with viscosity change $\Delta\mu = \mu_2-\mu_1 = -0.9\ \mu_1$. The slope parameter chosen was $a=4$. Nu was evaluated at the upper model edge.

Ra	Nu	
	constant μ	variable μ
38	1.0007	1.0007
39	1.0007	1.0007
40	1.0007	1.0007
41	1.0008	1.0013
42	1.0015	1.0035
45	1.0572	1.1028
60	1.6092	1.6342

Table 6.2: Nusselt numbers for constant and variable viscosity in dependence of Rayleigh numbers

In case of convection, single cells emerged. See Fig. 6.4 and Fig. 6.5 for streamlines and isotherm contours for constant and variable viscosity cases - both for $Ra=60$. The contour levels are chosen equidistant in order to get an impression of velocities by the 'density' of streamlines. Where streamlines are closer - as in Fig. 6.4 at the bottom and left (ascending) side - velocities are higher.

Onset of convection for both cases is almost identical. Nusselt numbers start to deviate from the constant level of 1.0007 at $Ra\approx41$. For the constant viscosity case value this coincides well with the value of $4\pi^2=39.5$, theoretically determined by Lapwood (1948) - see chapter 5. It must be taken into consideration for the variable μ case that Ra is evaluated with a reference viscosity which is not identical with the viscosity at the top of the model (surface viscosity). If surface viscosity is taken as reference viscosity another value of critical Ra will result ('surface' will be denoted by superscript 's'):

$$Ra_{crit}^{s} = Ra_{crit}\frac{\mu_{ref}}{\mu_1} = \frac{41}{5} = 8.1 \tag{6.18}$$

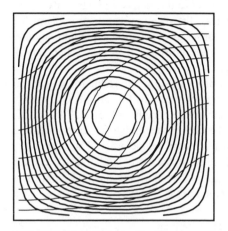
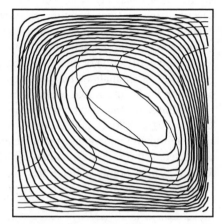

Fig. 6.4: Constant viscosity case   **Fig. 6.5:** Variable viscosity case
Closed curves are streamlines / others are lines of constant temperature

τ	$T_2(°C)$	a	μ_1/μ_2	$\mu_1-\mu_2$	θ_{ref}	$-\theta_0$	Ra_{crit}^{s} FAST-C	Ra_{crit}^{s} theory
0.34	119.7	5.722	0.230	0.770	1.167	1.536	17-17.5	17.25
0.50	166.7	3.890	0.162	0.838	0.796	1.045	12.5-13	13.35
0.67	216.5	2.904	0.123	0.877	0.592	0.779	9.5-10	10.76
0.84	266.3	2.317	0.101	0.899	0.475	0.623	7.5-8	8.86

Table 6.3: Critical Ra numbers in dependence of overheat capacity (T_1=25°C)

It can be stated: if Ra is defined by using surface viscosity, the critical value for the onset of convection is much smaller in the variable viscosity case than in the constant viscosity case. This phenomenon was predicted by Straus/Schubert (1977) using 1D-modeling. Higher dimensional cases were treated by Kassoy/Zebib (1976). They defined an *'overheat capacity'* parameter $\tau = (T_2-T_1)/T_1$ and determined values for Ra_{crit}^{s} in dependence on τ.

The last two columns of Table 6.3 compare numerical results from FAST-C(2D) with values theoretically derived by Kassoy (1975) and Kassoy/Zebib (1976). A more recent paper of Garg/Kassoy (1981) gives slightly differing values for τ=0.34 and τ=0.68; these values are $1.6\pi^2$=15.79 and π^2=9.81 respectively. There are intervals noted in the column for numerical results. The two interval limits are: (1) the highest value for Ra where no convection occurred; (2) the lowest Ra with emergence of a convection cell.

The coincidence between analytical and numerical results is remarkable, because not only the base set of differential equations is different, but Kassoy/Zebib (1976) solve an eigenvalue problem, while FAST-C(2D) simulates the transient behavior. Moreover, the temperature dependency of viscosity is not described by the same function. Nevertheless, the results given here confirm the general effect of a varying viscosity on the onset of convection - qualitatively and quantitatively. The theoretical derivations are confirmed. It is also a benchmark test for the verification of numerical codes.

A series of simulations has been run in order to show the reduction of critical limits for Ra for the whole temperature range between 20°C and 268°C (see Fig. 6.6). Minimum and maximum temperatures were taken out of {20°C, 60°C, 100°C, 140°C, 180°C, 220°C, 268°C}. The curves are drawn for cases with the same minimum temperature, but different maximum temperature. The curves are connected with the corresponding T_2=T_1 point at the horizontal line for critical Ra=$4\pi^2$. These points indicate the limit $\Delta\mu > 0$ in the presented diagram. The critical values shown in Table 6.2 are presented in the figure as well.

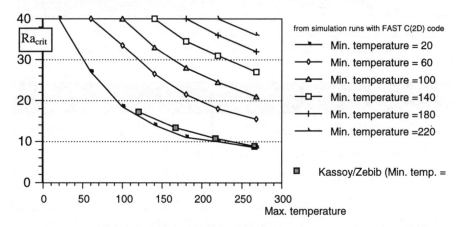

Fig. 6.6: Critical Rayleigh numbers for minimum/maximum temperatures between. 20°C and 268°C; all temperatures in °C

The following concerns the influence of variable viscosity on critical Rayleigh numbers in slender cavities. Chapter 6.1 shows that the critical margin increases significantly when the convection cell is slender (see Table 6.1). results from variable viscosity models are presented in Table 6.4. The two last columns of the table show the critical Rayleigh numbers which result from numerical modeling, when variable viscosity is taken into account. The numerical experiments have been carried out with FAST-C(2D) using the same 20×20 grid for a cavity with given length L and height H.

Minimum temperature in all simulations is 20°C. The viscosity has an tremendous effect: the critical margin calculated for the variable viscosity case is 4 to 6 times smaller than the value derived under the assumption of constant viscosity.

Aspect ratio H/L	Constant μ	T_{max}=100°C	T_{max}=268°C
1	39.5		9
2	61.5		12
4	179		30
8	652	300	125

Table 6.4: Onset of convection in slender cavities (Critical Rayleigh numbers in dependence of aspect ratio and maximum temperature)

6.2.3 Heat Transfer

One main concern of convection studies is, to find the connection between flow patterns and heat transfer. The total heat transfer through the porous medium from the high temperature at the bottom to the low temperature boundary at the top is described by Nu. Nu relates the heat transfer of the convection case to the heat transfer of the conduction case. In subcritical cases it is 1; in supercritical cases it always exceeds 1 (compare chapter 3.9).

Nu for steady state can generally be derived in an arbitrary horizontal layer of the system. In the interior two components of heat transfer have to be considered: advective and diffusive transport add for the total heat transfer. At the closed upper and lower boundary it is only diffusion is possible. In order to get Nu from the numerically calculated values of θ, it is sufficient to calculate the mean temperature gradient per unit length at the horizontal boundaries. The diffusivity is not required, because the increase relative to the conduction state has to be derived only.

If a linear approximation of the gradients is chosen, equation (4.38a) can be taken. One row of nodes in the interior of the system will then be used for the calculation of Nu. The following results used cubic approximation (4.38c) of the θ–gradients, for which the three lines of inner nodes closest to the horizontal boundaries have to be taken into account.

Nu depends on the convection pattern - in addition to the influence of Ra. The geometry of the whole system determines which convection patterns are possible. The lowest mode (1^{rst} mode) in a square cavity filled with a porous medium shows a 2D-eddy with equal length and height, i.e. with the cell aspect ratio length/height=1. If the system has aspect ratio 1 as well, the second possible mode (2nd mode) has two eddies with a cell aspect ratio of ½.

Nu for both cell systems have been determined from steady state numerical results with FAST-C(2D). Solution branches have been traced as noted above: once a cellular flow pattern had been obtained, the program has been restarted with a slightly altered value for *Ra* again and again.

The minimum temperature for all curves was set 20°C, while the maximum temperature has been varied as noted in Fig. 6.7 and Fig. 6.8. The plots show that a system with higher total temperature difference has higher heat transfer - for a fixed value of *Ra* and an unchanged convection mode.

A comparison of the constant and variable viscosity cases shows that heat transfer through the system is higher if changes in viscosity are taken into account. This finding seems to be important concerning the fact that values for *Nu* observed in experiments are in most cases higher than the ones predicted by analytical derivations or numerical simulations (see Combarnous/Bories (1975), Cheng (1978), Wang/Bejan (1987)). There have been various explanations for that effect. The results in this paper indicate another explanation: the variable viscosity effect. Analytical and numerical methods are mostly restricted to the constant viscosity case, but in real systems the variable viscosity effect can seldom be excluded.

The 2nd mode solutions for low *Ra* deliver a heat transfer lower than that of the 1[rst] mode. The situation is reversed for high *Ra*. Table 6.5 gives an overview on values of *Ra* and *Nu* for when the total heat transfer in both modes is equal.

It has to be noted that values for *Nu* summarized in the table are almost equal: between 3.2 and 3.35. Values for *Ra* slightly decrease with increasing T_{max}.

Nusselt-no.

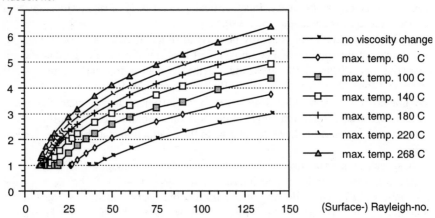

Fig. 6.7: Nusselt numbers calculated from numerical results for 1[rst] mode convection

Nusselt-no.

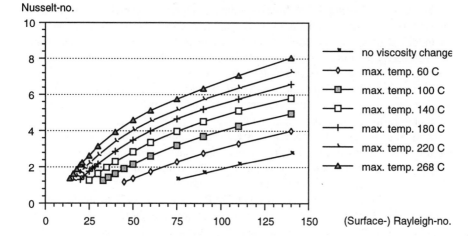

Fig. 6.8: Nusselt numbers calc. from numerical results for 2nd mode convection

The Platzman (1965) criterion, formulated for free fluids, states that the most likely convection mode is the one with the highest heat transfer. If that is true in the porous medium situation as well, it could be deduced from the data in Table 6.5 that 2nd mode convection has to be observed for $Nu > 3.42$ - provided that there is no other possible mode with higher heat transfer. With no further condition follows: when $Nu > 3.42$ no 1^{rst} mode convection should occur.

$T_2\,[^\circ C]$	const. μ	60	100	140	180	220	268
$Ra_{1,2}$	177	113	77	57	46	37	31
$Nu_{1,2}$	3.35-3.42	3.32-3.41	3.24-3.40	3.15-3.27	3.24-3.30	3.17-3.34	3.18-3.25

Table 6.5: Rayleigh- and Nusselt numbers for 1^{rst} and 2nd mode with equal heat transfer

6.3 Convection in Cold Groundwater

Most studies on convection deal with temperatures considerably above the freezing point. Low temperatures in most of the papers are above 20°C. Only few papers are dealing with lower temperatures, although this is the usual situation in near-surface groundwater in winter months under moderate climatic conditions, particularly in regions with colder climate.

Sun e.a. (1970) applied a linear stability analysis to investigate the onset of convection for liquids with a density maximum. For this case the Rayleigh number is modified using the coefficients in the cubic density-temperature relationship. Unfortunately no hints are given about the amount of change for some temperature intervals of interest. An experimental set-up is described by Yen (1974) and used to determine the relationship between Rayleigh and Nusselt number. A dependency on temperature range could be verified experimentally for temperature-intervals of more than 10°C.

A smaller temperature gradient (<8°C) was studied by Blake e.a. (1984). The authors use a quadratic density-temperature state equation and modify the Rayleigh number according to that function. The numerical analysis is similar to the one used here, but based on a constant viscosity.

Linear stability analysis is used again by Poulikakos (1985) to investigate the transient development of the system. The parameters are chosen in a way that the situation is supercritical and an initial perturbation develops into a stable convective pattern. This chapter uses the same approach in parts, but the focus remains on steady state.

Numerical experiments were used by Holzbecher (1997a) to obtain a complete picture of steady state convection for temperature intervals of different size, but with lower temperatures at 0°C, 2°C and 4°C. Viscosity changes were also taken into account. The numerical models, set up by the FAST-C(2D) code, were based on the streamfunction formulation, as described in the following chapter.

6.3.1 Streamfunction Formulation

The analytical formulation of convection is based on the principles of mass and energy conservation. If the Oberbeck-Boussinesq assumption is valid in the 2D case, the streamfunction Ψ can be introduced to describe the flow field. The following flow equation has already been presented as equation (3.28):

$$\left[\frac{\partial}{\partial x} \frac{\mu}{k_z} \frac{\partial}{\partial x} + \frac{\partial}{\partial z} \frac{\mu}{k_x} \frac{\partial}{\partial z} \right] \Psi = -g \frac{\partial \rho}{\partial x} \qquad (6.19)$$

If the porous medium is homogeneous and isotropic and, if viscosity is constant, a parameter transformation to dimensionless form leads to the following flow equation:

$$\nabla^2 \Psi = \frac{Ra}{\Delta \rho} \frac{\partial \rho}{\partial \theta} \frac{\partial \theta}{\partial x} \qquad (6.20)$$

If a constant Rayleigh number is further defined using the total change of density from the bottom to the top of the system, a correction factor f_ρ has to be introduced:

$$\nabla^2 \Psi = Ra \cdot f_\rho \frac{\partial \theta}{\partial x}$$

$$\text{with } f_\rho = \frac{1}{\Delta\rho}\frac{\partial\rho}{\partial\theta} = \frac{1}{\Delta\rho}\frac{\partial\rho}{\partial T}\frac{\partial T}{\partial\theta} = \frac{T_{max} - T_{min}}{\Delta\rho}\frac{\partial\rho}{\partial T}$$

(6.21)

Note that the correction factor changes with local temperature spatially and temporally. A different point of view may understand the product $Ra \cdot f_\rho$ as a local Rayleigh number.

In the same manner, a correction factor f_ρ is introduced to account for changing viscosity. A reference value for dynamic viscosity μ_{ref} must be selected to calculate the Rayleigh number as a constant characteristic of the system. The correction factor f_μ has already been introduced in chapter 6.2. Note that f_μ varies with temperature:

$$\nabla^2 \Psi = Ra \cdot f_\rho f_\mu \frac{\partial \theta}{\partial x} \qquad \text{with } f_\mu = \frac{\mu_{ref}}{\mu}$$

(6.22)

The viscosity value at the surface, i.e. for the lowest temperature, is chosen as the reference in correspondence with Kassoy e.a. (1975).

The system is completed by the transport equation (3.20b), by the defining equations for streamfunction (3.26), and by the equations of state which describe the temperature dependency of density and viscosity. Density changes as function of temperature between 0°C and 20°C are given in formulae (2.2) and (2.3). The well-known anomaly of water is expressed in that the graph has a maximum value at T=3.98°C. Fig. 2.3 depicts the relevant temperature dependency.

When the temperature dependency is given in a functional form $\rho(T)$, the correction factors can be calculated from these by use of formula (6.20). Fig. 6.9. gives a graphical representation. It should be noted that f_ρ is negative below 4°C and positive above 4°C. This formulation is implemented in the FAST-C(2D) code.

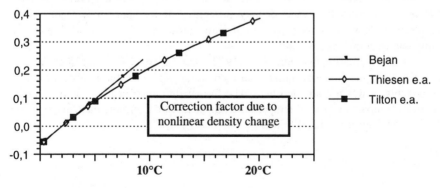

Fig. 6.9 : Rayleigh number correction factor for nonlinear density change

6.3.2 Onset of Convection

Holzbecher (1997a) considers different temperature intervals. Temperature ranges entirely above 4°C are initially studied. Some results are listed in Table 6.6. The numerical simulation confirms the expected result; the reduction of the critical Rayleigh number can mainly be attributed to variable viscosity.

Steady convective motions have been observed by Holzbecher (1997a), when the minimum temperature is below 4°C and the maximum temperature is above 12°C. But the critical Rayleigh numbers are further reduced, compared with the previously mentioned temperature range. Flow patterns are discussed in the following chapter.

The tendency of decreasing critical margin changes in systems having a maximum temperature at the bottom slightly above 4°C. Convective flow patterns in the unit square change and the critical Rayleigh number, detected by branch tracing in the numerical, even increases. The temperature interval [0°C,6°C] for example, has a critical margin of Ra_{crit}=240.

Minimum temperature [°C]	4	4	4	8	8	12
Maximum temperature [°C]	8	12	20	12	20	20
Critical Ra (linear ρ)	36.5	34.5	31.0	36.5	33.0	35.0
Critical Ra (nonlinear ρ)	35.5	32.5	28.5	36.5	32.0	34.5

Table 6.6: Critical Rayleigh numbers for temperature intervals above 4°C with and without consideration of nonlinear density changes

It has to be recognized that the density difference $\Delta\rho$, used in definition of the Rayleigh number (3.45), becomes irrelevant in the cold water case. Besides the fact that $\Delta\rho$ has the opposite sign in degenerate cases, in the temperature range of question it underestimates the density changes in the potentially unstable part. While $\Delta\rho$ denotes the water density difference at boundary temperatures, the convective flow is moved by the density difference between 4°C and the temperature at the lower boundary.

This is the reason that Sun e.a. (1970) and Bejan (1987) introduce a new definition for the Rayleigh number in cold water. Although the new definition is reasonable, it was not adopted here. The former definition has been kept in order to make comparisons for all temperature ranges and to determine the margin where the classical approach breaks down.

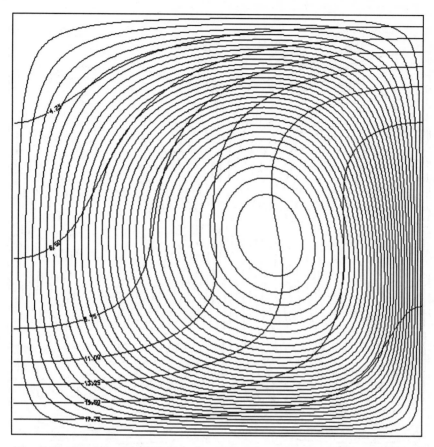

Fig. 6.10: Steady state convective flow patterns for cold water in porous medium
for T_{min}=2°C, T_{max}=20°(case A), calculated with FEFLOW (1996)[3]

6.3.3 Flow Patterns

It is well known from classical theory that the lowest unstable mode is two-
dimensional. If density changes are linear the corresponding eddies have equal
length and height, i.e. the aspect ratio H/L is 1 (compare chapter 5). For
temperatures above 25°C it has been shown by Kassoy/Zebib (1975) that the
patterns at the onset of convection remain almost the same, if viscosity changes are
taken into account. A small change in the corresponding wavenumber was detected
by Morland e.a. (1977) in the case of nonlinear density changes for temperatures
above 20°C. The situation for temperatures below 20°C is presented in the
following.

---

[3] Fig. 6.10 to Fig. 6.13 to with courtesy of H.-J.G. Diersch

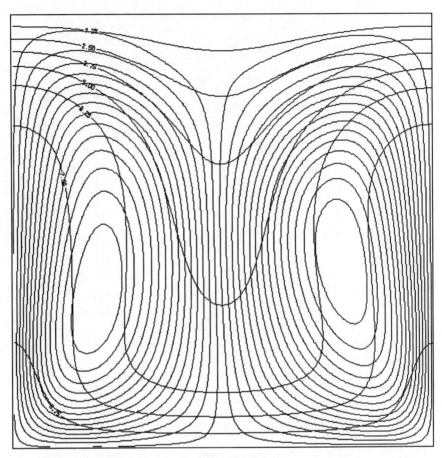

Fig. 6.11: Steady state convective flow patterns for cold water in porous medium for $T_{min}=0°C$, $T_{max}=10°$ (case B), calculated with FEFLOW (1996)

Convection patterns in cold water convection are different which has already been mentioned by Blake e.a. (1984) and Poulikakos (1985). At temperatures below 3.98° C there is a part of the model region near the upper boundary with a stable density gradient. This layer is not penetrated by convective motions as in the case of overall unstable density gradients. Eddies appear near the upper boundary in supercritical steady state flow.

The convection patterns for various temperature intervals presented by Holzbecher (1997a) have been basically confirmed by a calculation with FEFLOW (1996). Steady state patterns calculated by FEFLOW are shown in Figures 6.9 to 6.12.

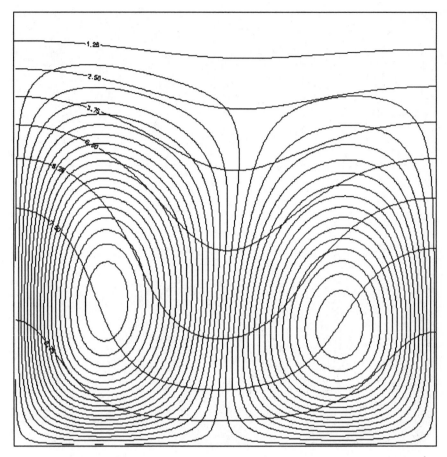

Fig. 6.12: Steady state convective flow patterns for cold water in porous medium for $T_{min}=0°C$, $T_{max}=10°$(case C), calculated with FEFLOW (1996)

Fig. 6.10 shows a 1-eddy pattern which looks very similar to the patterns shown by Kassoy/Zebib (1975). Narrow streamlines are observed in hot water, i.e. near the bottom and in ascending flow. Fig. 6.11 shows a 2-cell pattern with descending fluid in the center of the cavity. Fig. 6.12 shows two convection cells as well. The upper two cells are not visible in the streamfunction plots, which are drawn for a (too-) small number of equidistant levels. Eddies in the top part are much less pronounced than the eddies in the lower part. Fig. 6.13 gives a more illustrative view of the phenomenon. In the lower part of the system three strong circulating rolls appear. In the upper part eight small eddies are visible from the streamline contour plot. These small eddies are very weak. This can be derived from the fact that isotherms are horizontal; in the upper part heat conduction is the most relevant process.

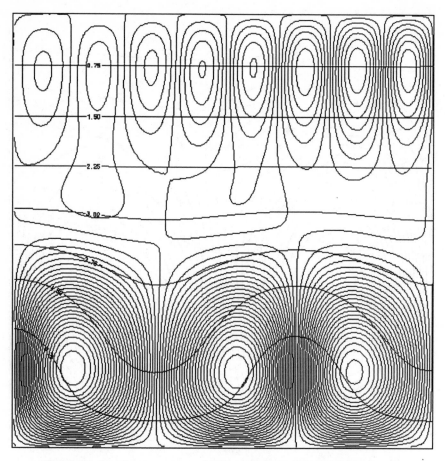

Fig. 6.13: Steady state convective flow patterns for cold water in porous medium for $T_{min}=0°C$, $T_{max}=6°$ (case D), calculated with FEFLOW (1996)

In the vicinity of the top boundary the horizontal layering of isotherms is only slightly disturbed. The vertical extension of this upper boundary layer can be attributed to the temperature interval: in Fig 6.11 the potentially unstable range extends over a temperature interval of 6°C, while the maximum temperature difference is 10°C. In Fig. 6.13 the potentially unstable range extends over 2°C, while the total temperature difference is 6°C.

Holzbecher (1997a) calculated convective motions for the same temperature. The Rayleigh number for runs represented in Fig. 6.10 and Fig. 6.11 is 60, 28.5 for Fig. 6.12 and 300 for Fig. 6.13. The correspondence with results calculated with FAST-C(2D) are remarkable for the flow patterns shown in Fig.s 6.10, 6.11 and 6.12. The rotation sense is different. This can probably be attributed to different small perturbations in the initial state. Concerning the steady state the rotation is irrelevant because of the symmetry of the system.

Property	Case A	Case B	Case C	Case D	Unit
Hydr. conductivity	54.31	675.8	321.02	4691.	10^{-4} m/s
Density expansion	$\rho=a+bT+cT^2+dT^3+eT^4+fT^5+gT^6$ with $a=999.8396$, $b=6.764771\cdot10^{-2}$, $c=-8.993699\cdot10^{-3}$ $d=9.143518\cdot10^{-5}$, $e=-8.907391\cdot10^{-7}$ $f=5.291959\cdot10^{-9}$, $g=-1.359813\cdot10^{-11}$				kg/m^3
Viscosity dependency	$\dfrac{\mu_{ref}}{\mu}=\dfrac{1+0.7063\zeta-0.04832\zeta^3}{1+0.7063\zeta_{(T=T_0)}-0.04832\zeta^3_{(T=T_0)}}$ with $\zeta=\dfrac{T-150}{100}$ at $T\,[^{\circ}C]$				
Thermal diffusivity	$1.5476\cdot10^{-7}$				m^2/s
Rayleigh number	61.05	61.05	29	304.4	-

Table 6.7: Input data for FEFLOW for cases A-D, shown in figures above

Flow patterns for the temperature interval [0°C,6°C], shown in Fig. 6.13, FEFLOW and FAST-C(2D) results are different. FEFLOW calculates three rolls in the lower part, while FAST-C(2D) results show only two rolls.

FEFLOW output has more rolls in the upper part. Holzbecher (1997a) reports only two eddies there what may be attributed to the higher resolution of the FE-grid used in FEFLOW calculations; the FAST-C(2D) model uses a coarse 20×20 grid. However, the differences in convection rolls are marginal in regions with dominating conductive regime.

6.4 Relevance of Convection in Natural Systems

Chapter 2 already mentions the geothermal gradient (increasing temperature with depth). It is caused by heat flow from the earth core. Heat from the interior of the earth principally produces the potential unstable situation of geological formation heated from below. The criterion with the critical Rayleigh number can be applied (see chapter 5.1) to decide, if convective flow can occur.

Bjørlykke e.a. (1988) assume typical values for most geological and fluid parameters and discuss the relevance of thermal convection in sedimentary basins. They come to the result that, for sand with the permeability of $k=1$ Darcy ($=9.8697\cdot10^{-13}$ m^2), the porous layer should exceed a height of 330 m in order to favor convective motions. Moreover, it is argued salt gradients dominate the density variation. According to Bjørlykke e.a. (1988) many measurements have shown that there '*is no overall density reduction with depth*'.

Some arguments should be added to that argumentation. Bjørlykke e.a. (1988) do not take into account the effect of anisotropy and of variable viscosity. As chapters 5.1 and 6.2 show both of these phenomena tend to reduce the critical Rayleigh number and thus the critical thickness of the formation. It should be recognized that, in the case of convection, the temperature increase with depth is very different depending on the location where the measurement is taken. As mentioned above, S-shaped curves for the isotherms occur in 2D convection cells (see Fig. 5.4). Some locations will have a small temperature gradient and others a large gradient. A single measurement in a convective system is hardly to be interpreted, if it is not known whether the borehole is placed in the downflow or in the upflow part of the sediment layer.

Phillips (1991) points to the fundamental difference of flow regimes in terrestrial aquifers on one side and submerged sediments on the other side. On the continents a hydraulic head gradient can generally be found. Even small gradients dominate buoyancy forces and density driven flow patterns can be observed under special conditions only. In submerged sediments the situation is different. Phillips (1991) writes: '*Pure thermally driven flows are generally to be found only when the hydraulic forcing is in effect absent, in submerged banks, beds, or continental shelf regions or in isolated, totally confined permeable strata*'. What is noted as 'thermally driven' is in fact density-driven flow. And density gradients may emerge from temperature or salinity gradients, as shown above. Flow in submersed porous media is density-driven.

Convective motions may appear in situations with an unstable salt gradient. In the vicinity of salt lakes or salt crusts this condition may be given. How convective motions may look like is discussed in chapter 14 about desert sedimentary basins.

Complex flow patterns may occur due to the combined effect of thermal and saline regime on density. These are only partially understood. Double diffusive convection (mostly for pre fluids) is discussed in Brandt/Fernando 1995).

Another argument concerns the boundary conditions. The assumption of isothermal conditions at both horizontal boundaries is rather arbitrary. A geophysical point of view holds latent heat flux from the interior of the earth to be the general process which determines temperature distribution in the earth crust. A mean temperature gradient corresponding to the mean heat flux could be used as condition at the horizontal boundaries. Numerical experiments with FAST-C(2D) and THERMOD (Springer 1995) showed that there is no steady state under these conditions. Palm/Tveitereid (1979) found that the critical value of the Rayleigh number is different when a perfect insulator boundary condition is introduced at the top boundary instead of the perfect conductor condition.

The symmetries play an important role for the bifurcation behavior of the mathematical solutions. Swift (1988) writes: '*in the presence of symmetry the typical bifurcation can be very complicated compared with systems without symmetry*'. Of course the symmetries of the idealized constellations, used for a mathematical treatment, are not given for geological formations in the field.

It may be concluded from all these arguments that an answer cannot easily be given to the question: is thermal convection a relevant processes in the sediments of the earth crust. The simplified situations discussed in this book are definitely not representative for real systems. However, they may be representative for some artificial systems in the laboratory or in the industry. And they are definitely good for testing and improving codes.

7 Oscillatory Convection

7.1 Hopf Bifurcation

Chapter 5.5 presented a sensitivity analysis concerning the change of steady convective motions with increasing Rayleigh number. The gradual increase of Ra was only possible up to a certain limit. When Ra was increased beyond that limit, there were difficulties in obtaining a steady situation. In figure 5.9 there is no value for the Nusselt number of the one-cell branch for $Ra=500$, because a steady state could not be reached in the numerical simulations.

Can this behavior be explained with difficulties in the numerical algorithm? Or is there a second critical Rayleigh number where the hydraulic flow regime changes again - as there is a sudden change from conduction to convection at the critical value of $4\pi^2$, which can be described by a pitchfork bifurcation (see chapter 5.2)? In fact some researchers have shown by that the second alternative is true: there is a second critical Rayleigh number Ra_{crit2}, at which the steady convective regime becomes unstable.

Straus (1974) used the spectral approach in combination with the Galerkin technique which had been previously applied for the pure fluid case by Veronis (1966). The approach is partially outlined in chapter 5.3. He found that, for a given value of Ra 'there is a range of horizontal wavenumbers for which stable two-dimensional convection exists'.

Another interesting result of Straus (1974) is that the extension of the stability-region is limited. There are obviously no stable 2D-convection cells for $Ra>380$. Straus (1974) restricted his study to 2D-patterns and suspects that, for higher Rayleigh numbers, 3D steady convective motions govern the system.

Caltagirone (1975), in addition to the mentioned Galerkin technique, uses numerical simulations to study the relation between Ra, Nu and the cell aspect ratio. The Galerkin technique leads him to a critical Rayleigh number of $Ra_{crit2}=384\pm5$ for the square convection cell. Numerical simulations with a 2D-system show there is a transition from the stable to the *fluctuating* convective regime.

Straus/Schubert (1979b) use the spectral method to treat 2D and 3D configurations and obtain a critical range for the change to unsteady motions for Ra between 300 and 320. The same authors later correct that result using improved numerical methods - with an increased number of terms in the approximating functions (Straus/Schubert 1982). Rayleigh numbers up to 650 are studied for the square 2D-cavity and several margins are found with transitions between different distinct unsteady states. Onset of oscillatory convection is given for Ra between 380 and 400. At a (third) critical value Ra_{crit3} between 480 and 500, a switch to a two frequency oscillatory quasi-periodic mode occurs. If Ra is increased further (500-520), a single frequency becomes dominant again - having a different frequency than those observed before.

Kimura e.a. (1986) give an update of previous results. Their calculations show the switch to a second single frequency periodic regime at Ra_{crit4} between 560 and 570. Kimura e.a. (1986) found the transition to chaotic regime for Rayleigh numbers between 850 and 1000 and write: '*routes to chaos in porous-medium convection are fundamentally different than those in ordinary fluids*'. They point to the fact that the step from a 2-freqency regime back to a single frequency regime (with increasing Ra) jumps from higher disorder to lower disorder; while in pure fluid convection there is a straight line with increasing disorder from conduction to chaos (Ruelle/Takens 1971, Bhattachargee 1987).

Caltagirone/Fabrie (1989) trace the dynamic regimes of unicellular convection in a square system up to $Ra=1500$ and basically confirm the previous findings. They also identify the emergence of a second, quasi-periodic pattern for Rayleigh numbers above 1000 (Ra_{crit5}). The change to chaos can be observed for higher values.

The references cited so far in this chapter use the spectral approach in combination with the Galerkin technique. The set of amplitude equations - a set of ordinary differential equations - is treated as initial value problem. There are some differences in the numerical treatment: Kimura e.a. (1986) and Caltagirone/Fabrie (1989) apply a modified technique, the '*pseudo-spectral*' method, with higher computational efficiency.

In contrast, the authors of the following publications use techniques from bifurcation theory to analyze the set of ordinary differential equations. This method has been applied in chapter 5.3 for a simpler system, to identify the pitchfork bifurcation at Ra_{crit}. The transition to oscillatory motions - at Ra_{crit2} - happens at a different type of bifurcation: the *Hopf bifurcation*.

Fig. 7.1 gives a graphical illustration of what takes place at this type of bifurcation (similar illustrations can be found in many textbooks, see Mullin 1993). The Hopf bifurcation describes the transition from a situation with an equilibrium point to a qualitatively different situation with a limit cycle. Point and circle represent the two different solution types in the phase space. Concerning convection, at a Hopf bifurcation there is a switch from a steady rotating state to an oscillatory rotating state.

A bifurcation appears when eigenvalues of the linear system cross the imaginary axis. When a pair of imaginary eigenvalues crosses the imaginary axis, a Hopf bifurcation may occur. For simple cases of bifurcating systems the critical value of a parameter at which the transition occurs, can be determined exactly by analytical means.

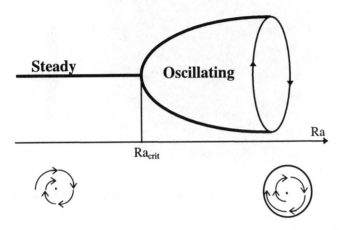

Fig. 7.1: Transition to oscillatory convection - illustration of the Hopf bifurcation

Aidun/Steen (1987) use an eigenvalue expansion and a branch-tracing technique to treat the bifurcation problem. They obtain $Ra_{crit2}=390.7$ for 2D problems. Aidun (1987) analyzes the 3D case and confirms the previous result, under the condition that the extension in the third dimension is small enough. Riley/Winters (1991) treat the problem by similar means and obtain a value of 390.72.

The analysis in chapter 5.3 calculates stable convection solutions, but is based on a very small series of modes (equation (5.24)). The result is thus of limited validity.

7.2 Simulation

FAST-C(2D) has been used to set up a the 20×20 block model for the 2D unit square[1]. Starting from a one-eddy flow pattern the unsteady development of the flow system was simulated. Initial state of model runs for $Ra=360$, $Ra=380$ and $Ra=400$ was the steady state one-cell solution, which has been calculated for $Ra=100$ (Fig. 7.2).

---

[1] The input data for the model with Ra=400 can be found on CD-ROM in file *Osc_Conv.WTX*

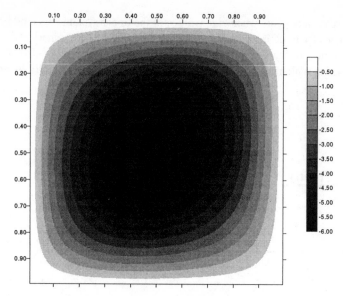

Fig. 7.2: Streamfunction distribution at initial state (steady state for Ra=100)

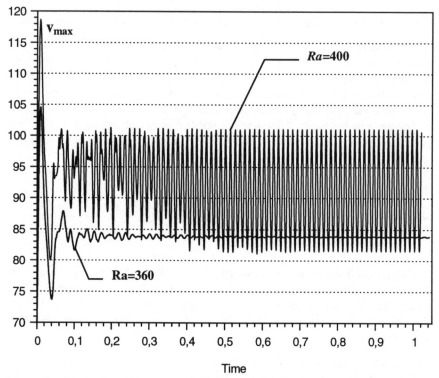

Fig. 7.3: Max velocity as function of time for simulation with Ra=360 and Ra=400

The model was run for a 20×20 grid in a square box with spatial extension $H=L=1$. The discretization type for 1st order derivatives in the transport equation was CIS (central in space). Timesteps were calculated automatically to fulfill the most restrictive of both Neumann- and Courant-criterion (timestepping parameter $\chi=1$ - compare chapter 4). Simulation was stopped at dimensionless time $t_{max} = 1$.

The numerical steady state was not reached in the simulations; i.e. there were still some changes in the state variables, although the system for $Ra=360$ seems to be steady state. The following figures visualize the transient development on velocity and streamfunction extrema.

For $Ra=360$ and 400 it is shown in Fig. 7.3, how the maximum of velocity changes with time. The difference between the two simulations is striking: while for the lower Rayleigh number a constant value is approached, the regime for the higher Rayleigh number is clearly oscillatory - For $Ra=400$ the numerical simulation reveals an oscillatory flow regime. For $Ra=360$ there are fluctuations as well. These are much less pronounced as for the other Rayleigh number and they show a decreasing amplitude.

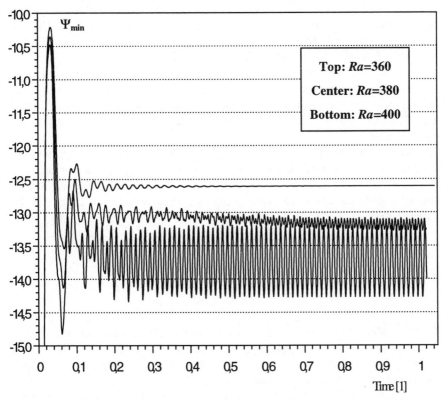

Fig. 7.4: Transient development of convection visualized by a time-series of Ψ_{min} for $Ra=360$ (high values), 380 (intermediate), 400 (low values)

Apparently the numerical model reproduces nicely the qualitative change of the solution proposed by the authors above mentioned. The system turns from steady state convection to oscillatory convection. The second critical Rayleigh number has to be sought between the values $Ra=360$ and $Ra=400$.

In all simulations reported so far the convection mode remained the same. Starting from the one cell situation in the square cross-section, there remained one cell for all Rayleigh numbers in the simulated time-period. This phenomenon was checked by a single simulation for $Ra=500$. Under that condition the flow pattern changed indeed: a two cell pattern emerges out of the one cell initial condition.

The minimum of the streamfunction can be found in the center of the convection eddies, while Ψ approaches zero near all boundaries. If Ψ_{min} is plotted against time, graphs as shown in Fig. 7.4 result. A steady state is approached for $Ra=360$: oscillations around the steady solution vanish. An oscillating solution is reached for $Ra=400$ - if plotted in a phase diagram a limit cycle would be approached (see Fig. 7.1 lower part). Neither a steady state nor a limit cycle is visible for $Ra=380$. There are oscillations, but there is no constant cycle: Ψ-values still tend to decline.

In the oscillating convection case (for $Ra=400$) the timeperiod of the oscillations is approximately 0.0128 in dimensionless time-units (see Fig. 7.6). The frequency is around 491. This correspondents reasonably well with the value of 471 derived by Riley/Winters (1991) for Ra=400.2.

The change of the flow pattern can best be viewed in an animation sequence.[2] The flow switches between two types of unicellular regimes. In one the flow in the interior away from the boundaries is almost circular. In the other extreme an elliptical shape of fluid flow can be observed. The ellipse is lying diagonally in the square system.

Note that the flow direction is not unique. If there is a convection cell with upward flow on the right side and downward flow on the left side, another cell with upward flow on the left side and downward flow on the right side is a valid solution as well. In the described example the rotation sense is given already in the initial condition.

7.3 Influence of Numerical Parameters

The results shown above are calculated with the CIS-approximation for the 1st order derivatives in the transport equation (see equation (4.5)). The BIS option (see equation (4.4)) has been selected in a further run for purposes of comparison. Results for $Ra=400$ are contrasted in Fig. 7.5.

The result is striking again. The oscillating regime cannot be determined, if the upwind scheme is used. The numerical simulation stops at the numerical steady

---

[2] The sequence can be viewed on CD-ROM in file *FASTdemo.HTML* or in the Internet

state, when the variables do not change within the chosen timestep. This situation depends on the selected timestep Δt and the accuracy ε. In the simulation, shown in Fig. 7.5, the timestep was chosen to match with the Courant- and Neumann-conditions and to fulfill the more restrictive exactly. The accuracy was chosen as 10^{-5}.

In the BIS simulation, shown in Fig. 7.5, the initial peaks are reduced but still visible. The fluctuations of the oscillatory regime are totally smoothed out. The system seems to have reached a steady state. Numerical dispersion (see equation (4.7)) is the explanation for the smoothing effect. The effect is so strong that even for Rayleigh numbers, much higher than the critical margin of $Ra_{crit2}=391$, no fluctuating regime can be observed.

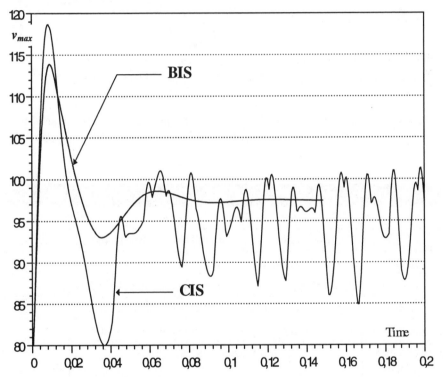

Fig. 7.5: v_{max} timeseries obtained by using BIS or CIS approximations for 1st order derivatives ($Ra=400$)

The dependence on several other numerical parameters has been studied for the reference case with $Ra=400$. The solvers and their precision option was changed. It made no difference for results, if the solvers were run with the single or the double precision option. The results were identical on the first six digits even after 1845 timesteps - when the specified end time ($t_{max}=1.$) was reached.

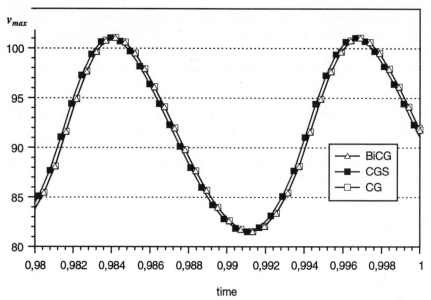

time

Fig. 7.6: Maxima of velocity in last simulated time period (Ra=400), obtained with different solution methods. There are four timesteps used to proceed from one marker to the next.

Denotation	Postprocessed item	Formula
Nu 1-up	1st order approx. of Nusselt number at upper boundary	
Nu 1-down	1st order approx. of Nusselt number at lower boundary	
Nu 2-up	2nd order approx. of Nusselt number at upper boundary	
Nu 2-down	2nd order approx. of Nusselt number at lower boundary	
Nu 3-up	3rd order approx. of Nusselt number at upper boundary	
Nu 3-down	3rd order approx. of Nusselt number at lower boundary	
Norm. max. velocity	Normalized max. velocity	$v_{max}/\max(v_{max})$
Norm. min. streamfunc.	Normalized minimum streamfunction	$\Psi_{min}/\min(\Psi_{min})$
Theta left	Medium of transport variable at left boundary	θ_{left}
Theta right	Medium of transport variable at right boundary	θ_{right}

Table 7.1: Post-processed items to study the transient development in oscillating solutions

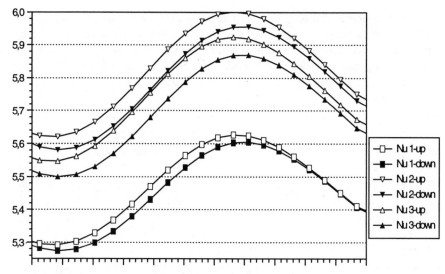

Fig. 7.7: Transient development of Nusselt numbers in one period
(see Table 7.1; 5 timesteps needed to proceed from one marker to the next)

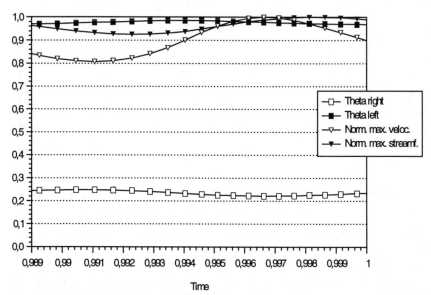

Fig. 7.8: Transient development in one period visualized on some selected items
(see Table 7.1; 5 timesteps needed to proceed from one marker to the next)

Minor differences could be observed when the solver type was changed. Differences for the iterative solvers occur because the solutions may differ behind a certain number of digits. The reason is that the criterion causing the iteration to stop is based on the residual and is fulfilled for a certain range of vectors for the unknown variables. The program was run using CG, BiCG and CGS as solution routines. The deviations are so small that they cannot be recognized on the timescale used above. That is the reason why only the last timeperiod is shown in Fig. 7.6. A small phase-shift can be observed in the results of the CGS-method, while CG and BiCG show exactly the same curves (marker for BiCG always lies behind the CG marker).

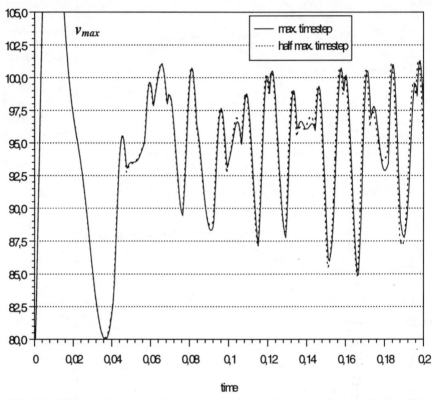

Fig. 7.9: Effects of timestep size; maximum velocities in simulations for $Ra=400$ at the beginning of the simulation period

There was also a check of how different items derived from the calculated variables reflect the oscillatory behavior of the system. An overview on the items is given in Table 7.1.

Maximum velocity and minimum streamfunction are normalized with the extreme value they receive within the oscillatory regime at the end of the simulated time ($t_{max} \approx 1.0$). Their maximum values become 1.0. The development of these items in the last simulated period is depicted in Fig. 7.8. The values obtained for Nusselt numbers depend very much on the order of approximation used for their calculation from the array of the transport variable. If first order approximation is selected, the Nu's in the time-period become quite low (see Table 7.2). Nu's are highest for quadratic approximation and intermediate for third order approximation. Nusselt numbers for upper and lower boundary are never equal. Nevertheless, minima and maxima of *Nu* are observed at the same simulated time.

	Nu 1-up	*Nu* 1-down	*Nu* 2-up	*Nu* 2-down	*Nu* 3-up	*Nu* 3-down
Max.	5.6269	5.6033	6.0014	5.9553	5.9241	5.8685
Min.	5.2921	5.2733	5.6205	5.5808	5.5471	5.4992

Table 7.2: Max. and min. of Nusselt numbers in the oscillation, depending on boundary and approximation order (Nu 1=1st order, see chapter 4;...)

The medium values of the normalized transport variable change only slightly as shown in the lower part of Fig. 7.8. The extreme values for velocity and stream-function are better indicators for the oscillatory behavior of the system. It should be noted that the extrema of Ψ_{min}, of v_{max} and of the Nusselt numbers can not be observed at the same time. The effect of timestep-size and timestep-selection has been checked additionally.

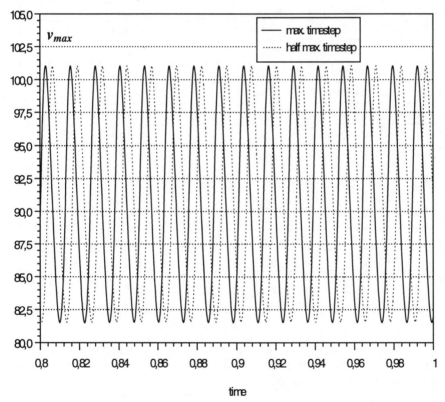

Fig. 7.10: Effects of timestep size; maximum velocities in simulations for *Ra*=400 at the end of the simulation period

Parameter	Coarse grid (20 x 20), CGS		Fine grid (40 x 40)	
	max. Δt	half max. Δt	BiCG, max. Δt	CGS, half max. Δt
Max. Nu (1.order above)	5.6283		5.3385	
Max. Nu (1.order below)	5.6054		5.3339	
Min. Nu (1.order above)	5.2920		5.0544	
Min. Nu (1.order below)	5.2726		5.0547	
Med. Nu (1.order above)	5.4665		5.1762	
Med. Nu (1.order below)	5.4476		5.1762	
Max. Nu (2.order above)	6.0029		5.4206	
Max. Nu (2.order below)	5.9564		5.4152	
Min. Nu (2.order above)	5.6204		5.1303	
Min. Nu (2.order below)	5.5799		5.1300	
Med. Nu (2.order above)	5.8187		5.2537	
Med. Nu (2.order below)	5.7780		5.2533	
Max. Nu (3.order above)	5.9256		5.3715	
Max. Nu (3.order below)	5.8699		5.3664	
Min. Nu (3.order above)	5.5470		5.0840	
Min. Nu (3.order below)	5.4983		5.0839	
Med. Nu (3.order above)	5.7430		5.2067	
Med. Nu (3.order below)	5.6938		5.2064	
Min. Ψ_{min}	-14.2725	-14.2700	-14.0129	-14.0146
Max. Ψ_{min}	-13.1998	-13.1994	-13.3388	-13.3408
Med. Ψ_{min}	-13.7697	-13.7663	-13.7179	-13.7211
Max. v_{max}	101.0383	101.0575	107.3761	107.3796
Min. v_{max}	81.4892	81.4904	93.2898	93.0247
Med. v_{max}	91.2532	91.2897	102.7907	102.7987

Table 7.3: Dependence of post-processed items concerning the oscillating regime on grid refinement

The reference case with $Ra=400$ was run with the option to fulfill the more restrictive of Courant- and Neumann-criterion exactly, i.e. that one of the following equations with $\chi=1$ becomes true:

$$Cou = \frac{v\Delta t}{\Delta x} = \chi \qquad \text{or} \qquad Neu = 2\frac{D\Delta t}{(\Delta x)^2} = \chi \qquad (7.1)$$

This is the maximum timestep, a modeler can use if neither criterion is to be violated. A second run was started where the most restrictive of the conditions is fulfilled with $\chi=0.5$. The output of both simulations are visually compared in Fig. 7.9 and Fig. 7.10. The runs are denoted by '*max. timestep*' for the reference case and '*half max. timestep*'.

Differences in both simulations appear gradually. It can be observed that major differences occur near turning points where the wide timestep run is not able to resolve the details of the changing system. Nevertheless, as Fig. 7.10 shows, the propagation of errors only results in a phase shift. Amplitude and frequency of the oscillating regime are equally simulated in both runs. The general conclusion is that timestepping has no effect on the general type of solution, as long as the conditions are fulfilled.

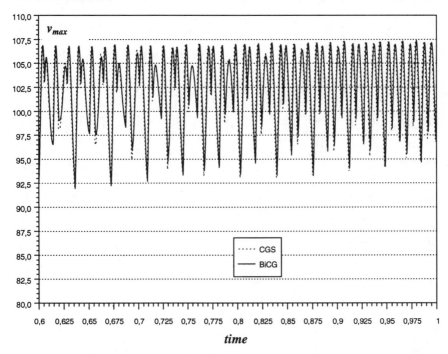

Fig. 7.11: Maximum velocity as indicator for thermal convection in a refined grid of 40×40 blocks for $Ra=400$ using two different CG solvers

Fig. 7.11 illustrates the oscillating regime in a refined grid. The results have been obtained for a 40×40 regular block system for $Ra=400$. Two different solvers have been used (CGS and BiCG) and an accuracy of $\varepsilon=10^{-5}$. The automatic timestepping parameter was chosen to 0.5 in the CGS run and to 1.0 in the BiCG run. Only the final parts of the simulations are shown in Fig. 7.11. The CGS run needed 8125 timesteps to proceed from initial time 0.0 to (dimensionless) end time 1.0; in the BiCG run 4076 timesteps were sufficient.

An oscillating regime is obtained in the fine grid simulations. Nevertheless, oscillations are not as regular as in runs with the coarse grid. Differences in the solutions with different CG-solver and different timestep are only marginal. There are differences between fine grid and coarse grid results which are summarized in Table 7.3. The presented values are minima, maxima and medium values of the simulated results in the timeperiod between $t=0.75$ and final $t_{max}=1.0$, a timeperiod with oscillating regime.

Nusselt numbers have been calculated in each timestep for upper and lower boundaries using three different approximation schemes (1st order, 2nd order and 3rd order). Minimum of streamfunction can be found in the center of the eddy. Maximum velocity is found near the edges of the system.

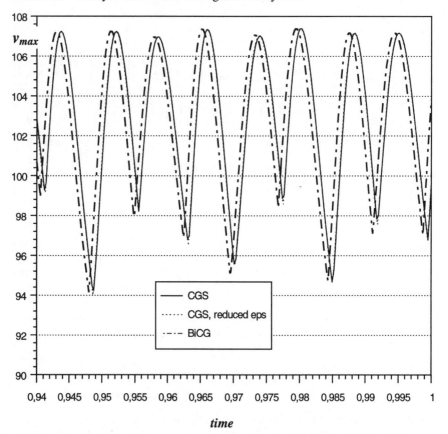

Fig. 7.12: Max. velocity for runs ($Ra=400$, 40×40 grid) with different solver options: the accuracy criterion was usually with $\varepsilon=10^{-5}$; the simulation with reduced value was done with $\varepsilon=10^{-6}$.

The spatial discretization has an tremendous effect on the results. Comparison of the results shows that Nusselt numbers are all overestimated in the coarse grid simulations. The differences are in the range between 0.3 and 0.6. If the stream-function extreme on the grid represents the strength of the inner eddy, it can be concluded that the oscillations in eddy strength are smaller when a fine grid is used. Nevertheless the mean value is almost the same. The maximum velocity, an indicator of eddy strength near the boundaries, is clearly underestimated in the coarse grid runs. This behavior is quite reasonable: grid refinement resolves the flow pattern relatively better where gradients are steep, i.e. in the outer parts, where the velocities are highest.

The output from the same simulation runs for a smaller timeperiod is illustrated in Fig. 7.12. The outcome from a simulation with reduced accuracy within the iterative CG solver is added. That run used $\varepsilon=10^{-6}$ instead of $\varepsilon=10^{-5}$. The same value for ε was taken for the residuals within the iterations to solve the linear equations and for the change of the unknown vector in the Picard iteration.

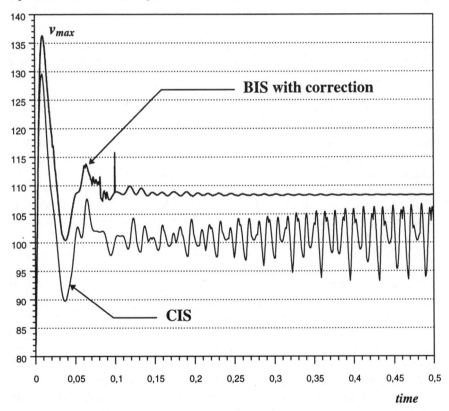

Fig. 7.13: Comparison of CIS and corrected BIS discretizations; maximum veloci-ties as function of simulated time (40×40 grid, Ra=400)

The picture portrays that the main difference in the results is a phase shift. This occurs as an effect of error propagation after several 1000 timesteps. Note that

BiCG results with usual ε and CG results with reduced ε are hardly distinguishable. Both have been applied with the max. timestep selection. The phase shift is thus probably due to the changed timestep option in the CGS-indicated run. Vice versa: there are almost no effects from the solution procedure, neither from solver type nor from prescribed accuracy. (Note: double precision option was used for all simulations on the 40×40 grid).

Another interesting aspect is given by a look at the grid-Péclet-number. Using v_{max} the maximum value for grid Péclet number can be estimated as

$$Pe_{max} = \frac{v_{max} \Delta x}{D} \qquad (7.2)$$

For dimensionless variables the expression becomes simply $v_{max}\Delta x$, because time scale is normalized using diffusivity in the transformation (see chapter 3.7). As v_{max} is in the range of 100 (see Table 7.3), the grid spacing needs to come down to 1/50, to fulfill the grid-Péclet-criterion in the entire system. Even in the fine grid discretization that is presented here the grid Péclet-criterion is not fulfilled.

The Péclet-criterion is only violated in the outer blocks of the system. Highest velocities can be observed in the vicinity of the boundaries. The behavior of the code was checked by applying the BIS discretization with truncation error correction (see chapter 4.1). Note that in those parts of the region, where the criterion is not fulfilled, the correction produces negative diffusivity. This should reduce numerical dispersion dramatically. Despite of this option, the results show that oscillations are suppressed which clearly appear in modeling with the CIS method. Oscillatory regime could not be observed when the BIS with correction was applied (see Fig. 7.13).

The fact that the grid Péclet criterion is not fulfilled, obviously causes problems in the numerical solution. Negative diffusivity in parts of the system does not eliminate numerical dispersion.

Finally it should be noted that all presented results are studied on unicellular motions in square cells. These type of cells have been chosen in order to compare with findings published by other researchers. These patterns are not even likely to exist in controlled experiments when the Rayleigh number is significantly above the critical value. Holzbecher (1991a) studied the emergence of various flow patterns in dependence of the Rayleigh-number. In fact it turns out that the stability of one-eddy flow regimes strongly depends on the unicellular initial condition shown in Fig. 7.2.

For comparison the same model was run with a different initial condition. The layered temperature distribution in a no-flow regime was disturbed at a single location. Starting from this initial condition a three-eddy steady pattern was approached. This numerical experiment indicates that unicellular motions can only be reached when the initial state is near the 1-eddy branch. Thus it can be questioned whether oscillatory convection is of any relevance beyond theoretical investigations.

Few studies on oscillatory convection in porous media can be found in the scientific literature. Oscillatory motions have been observed in laboratory experiments. Analysis using bifurcation theory based on several simplifying assumptions predicts the onset of oscillations when a parameter exceeds a certain critical margin. The best result that the research can deliver is that experimental and analytical results agree. But is this result of any practical relevance? It seems to be not. At least no such findings are known to the author.

Nevertheless computer models are useful because they allow to get insight in a specific problem by numerical experiments. Once a model is set up this is easy to do. Compared with laboratory experiments this is a cheap alternative. Initial or boundary conditions can be changed. Homogeneous distributions of parameters can be replaced by inhomogeneous distributions. Isotropic and anisotropic situations can be examined. Geometric parameters can be changed. In this chapter few of these possibilities have been used.

By numerical experiments experience with the behavior of nonlinear systems can be obtained. It is another task to check the relevance of numerical results for laboratory or field situations. But the direction and guideline for such research can be started by using the inexpensive tool of numerical modeling beforehand.

There is another argument for modeling such situations as oscillatory convection which do not have practical relevance yet: they can be used in a verification exercise for numerical codes. In fact this was the argument that has motivated the author to make the examinations presented in this chapter.

8 Horizontal Heat and Mass Transfer

The case of vertical heat and mass transfer has attracted a lot of researchers over decades. Its popularity can be attributed to the fascinating concept of a critical value of a dimensionless parameter combination, for which it is one of the illustrative examples. The classical example from fluid mechanics is the qualitative change from laminar to turbulent flow at the critical Reynolds number. The critical Rayleigh number characterizes the margin between conductive and convective regimes in free fluids and in porous media flow (see chapters 5 to 7).

Much less scientific attention has been given to the case where the situation is turned by an angle of 90°. Natural and man-made structures often have a horizontal gradient of heat or salinity. Buildings give the situation in the insulation of walls. Cold- or hot-storage installations are relevant not only in industry. Particularly seasonal energy storage facilities of either hot or cold fluid may be placed in subsurface porous media. Cryogenics and chromatography have interests in horizontal gradients of heat and gradients of solute concentration.

8.1 Analytical Approximations and Solutions

8.1.1 Convection

Podtschaske (1985) studied convective motions in horizontal tubes filled with a porous medium for another purpose. The analysis was done within the framework German PSE (Project Safety Assessment for a Nuclear Waste Repository). The scenario concerns the release of radionuclides after filling the underground passage between mine chambers with porous material. There may appear temperature and/or salinity gradients from one chamber to the other - in Germany the repository for high level waste is planned in the interior of a salt-dome (see chapter 13). If a convective motion emerges it may be easier for radionuclides to pass the porous medium barrier.

Podtschaske (1985) makes several simplifying assumptions. Concerning the geometry it should be allowed to treat the tunnel as a long thin rectangular cross-section. A further assumption is that only one of the gradients, of temperature or of salinity, from one vertical side of the tube to the other is different.

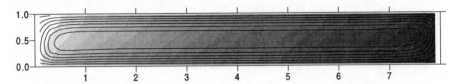

Fig. 8.1: Horizontal heat or mass transfer; flow is depicted by streamlines; temperatures or salt concentrations are depicted by gray patterns - calculated with FAST-C(2D) for dimensionless variables for a regular 40×21 grid

In an analytical treatment of the problem Podtschaske (1985) makes the additional assumption that the tube is of infinite length. For a 2D-geometry and unit height $H=1$, the following formula can be derived for the horizontal dimensionless velocities

$$v_x = R\bar{a}\left(z - \frac{1}{2}\right) \qquad \text{with } R\bar{a} = \frac{kg\beta \frac{\partial T}{\partial x} \rho H^2}{\mu D} \qquad (8.1)$$

The gradient $\partial T/\partial x$ is assumed to be constant. Note that the Rayleigh number definition is not consistent with the formulation given by Lapwood (1948). It is possible to reformulate Podtschaske's analysis using the definition of Ra from chapter 3. The following will show that this leads to a different formulation than given in (8.1). Starting with the differential equation (3.45) one may assume that flow is horizontal almost everywhere:

$$\frac{\partial^2 \Psi}{\partial z^2} = Ra \frac{\partial \theta}{\partial x} \qquad (8.2)$$

Further consideration of the defining relation $v_x = -\partial\Psi/\partial z$ (see (3.26b)) implies:

$$\frac{\partial}{\partial z} v_x = -Ra \frac{\partial \theta}{\partial x} \qquad (8.3)$$

By integration over z the result can be obtained that v_x is a linear function of z:

$$v_x = -Ra \frac{\partial \theta}{\partial x} (z - z_0) \qquad (8.4)$$

This procedure is allowed under the assumption that $\partial\theta/\partial x$ does not depend on z. The integration constant z_0 is necessarily equal to ½, because of the symmetry of the system: in the middle of the tube holds $v_x=0$. Note that the (8.2) - (8.4) hold in the dimensionless formulation as well. Nevertheless an analogous derivation can be made using variables with physical units.

Similar approximations for the horizontal component of the convective flow have been published by Bejan/Tien (1978) and Walker/Homsy (1978). Nield/Bejan (1992) distinguish between shallow layer and boundary layer regimes. The shallow layer ($L{>>}H$) may have finite length. Then the gradient of the transport variable may be approximated by the difference of θ given at the two vertical boundaries:

$$v_x = -Ra\frac{\theta_{max} - \theta_{min}}{L}(z - z_0)$$ (8.5)

Boundary conditions are of Dirichlet-type: θ_{max} on one, θ_{min} on the other side. Note that the sign of the velocity is arbitrary: the rotation direction of the convection roll is not specified by the above analysis. Numerical modeling can determine for which aspect ratios H/L the approximation holds (see 8.2.2).

Numerical simulations show also that stable convection solution exist for all Rayleigh numbers. A critical margin for the onset of convection does not exist. Roughly speaking, one could say that $Ra=0$ is the critical value: if there are any density gradients in horizontal direction (i.e. $Ra>0$), there will be convection.

The boundary layer case ($H{>>}L$) is more complex. The streamfunction equation can be simplified in analogy to the shallow layer case:

$$\frac{\partial^2 \Psi}{\partial x^2} = Ra\frac{\partial \theta}{\partial x}$$ (8.6)

But the transport equation cannot be reduced to an assumption of a constant gradient. Weber (1975) takes into account that the diffusion in vertical direction can be neglected. Using this for a steady state condition, equation (3.41) becomes simply:

$$\frac{\partial^2 \theta}{\partial x^2} = \frac{\partial \Psi}{\partial z}\frac{\partial \theta}{\partial x} - \frac{\partial \Psi}{\partial x}\frac{\partial \theta}{\partial z}$$ (8.7)

Weber (1975) uses an equivalent system to (8.6) and (8.7) and some further assumptions to derive an analytical solution for the unknown variables and an approximative rule for the Nu-Ra relationship:

$$Nu = 0.577\frac{L}{H}\sqrt{Ra}$$ (8.8)

Bejan/Tien (1978) tackle horizontal heat transfer in tubes of different shape.

8.1.2 Conduction

The statement above can be inverted: the pure conduction case is only possible, if there are no density differences. This point of view places pure conduction outside the scope of this book. Nevertheless, the pure conduction situation is of interest as one limit case of systems with weak convection. These systems are suitable test-cases for codes that are designed to treat variable density problems. It is usually possible to obtain a description of the conductive state by setting $\Delta\rho=0$ in the model input.

Häfner e.a. (1992) provide an extensive treatment of analytical solutions for the 1D heat conduction case. The situation, in which a system reacts to a sudden change of temperatures at the boundaries at time $t=0$ is described by the following initial and boundary conditions:

$$\theta(x,t = 0) = 0 \dots\dots\dots\dots\dots\dots\dots\dots\dots\dots(\text{initial condition})$$

$$\left.\begin{array}{l} \theta(x = 0, t > 0) = \theta_0 \\ \theta(x = L, t > 0) = \theta_1 \end{array}\right\} \dots\dots\dots\dots\dots(\text{boundary conditions}) \quad (8.9)$$

The description of conduction by a diffusivity D results in the following analytical solution for the transient development:

$$\theta(x,t) = \theta_0\frac{L-x}{L} + \theta_1\frac{x}{L} - 2\sum_{i=1}^{\infty}\frac{\theta_0 - (-1)^i\theta_1}{\pi i}\sin(\frac{i\pi x}{L})\exp(\frac{-i^2\pi^2 Dt}{L^2}) \quad (8.10)$$

A graphical representation of this solution with the dimensionless timescale Dt/L^2 is given below in Fig. 8.2. The solution is derived by Laplace- or Fourier-transformations (Häfner e.a. 1992) and holds for the case of cartesian coordinates. Radial cases as well as different types of boundary condition are shown by Häfner e.a. (1992).

8.2 Numerical Experiments

8.2.1 Conduction

Cartesian Coordinates

The 1D-situation described by the boundary and initial conditions (8.9) is simulated using FAST-C(2D) code. Diffusivity D is chosen as 1.0, the length L as well. Timesteps are 0.005 for the entire simulation period from t=0 to t=0.5. The Crank-Nicolson scheme the produces results as presented in the following figure.

The vertical axis shows normalized temperature. Prescribed temperature on the left boundary is set to θ_l=1; prescribed temperature on the right boundary is set to θ_2=2; initial temperature is set θ_0=0. Fig. 8.2 shows how the system gradually approaches the steady state, which is given by the straight line between 1 on the left and 2 on the right side. The timescale is transformed in dimensionless units by $t \rightarrow t \cdot D/L^2$. Note that the characteristic length in this problem is L in contrast to convection cases treated above (compare transformations (3.40)).

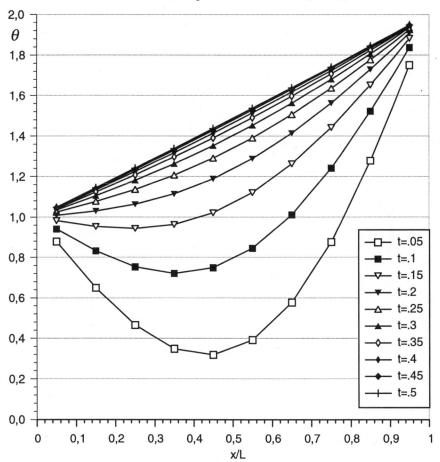

Fig. 8.2: Numerical and analytical solution for 1D conduction in cartesian coordinates for the case of cooling and heating from the side

The computations described below are a test for the FAST-C(2D) code. Numerical results are checked by comparison with the analytical solution given by formula (8.10). This is a feedback loop between the level of numerics and the level of analysis as outlined in chapter 1.2.

Fig. 8.2 shows the analytical solutions by lines and the numerical solutions by markers. Deviations between both are hardly distinguishable by view.

Radial Coordinates

Fig. 8.3 shows the same situation for the radial case. Note that the results depend on the inner and outer radius of the system (here denoted r_0 and r_0+L).

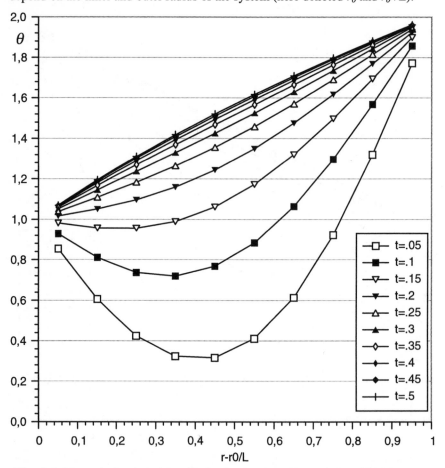

Fig. 8.3: Numerical solution for 1D conduction in radial coordinates for the case of cooling and heating from the side

In the cartesian coordinate system, the steady state of heat conduction between two boundaries with fixed temperatures is a linear curve (Fig. 8.2). In radial coordinates this simple relationship holds no longer. There the temperature gradient increases when the inner radius is approached. The gradient decreases towards the outer radius.

A numerical model set up with FAST-C(2D) calculates the quantity of this effect. Results are presented in Fig. 8.4, where the ratio of the inner radius r_0 and the horizontal length L is changed. The transient system is simulated in a system with 20 blocks in radial direction. The outcome of a simulation has been used as initial value for the following simulation with increased L/r_0 ratio.

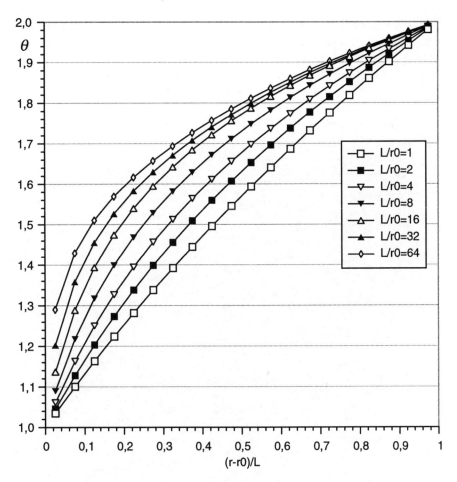

Fig. 8.4: Steady state of heat conduction problem in radial coordinates with varying ratio of L/r_0

8.2.2 Convection

In chapter 8.1.1 a linear relation between the horizontal velocity component v_x and the vertical coordinate z is derived for systems with $L>>H$. A model was set up to check the simple relation. A series of model runs was started using the steady state option in a 50×50 grid model. Rayleigh number Ra and system length L were varied. The results can be transferred to systems with arbitrary height H and length $H \cdot L$.

Fig. 8.5 and Fig. 8.6 show the streamfunction and transport variable obtained with FAST-C(2D) code for $Ra=100$ and for $H/L=1$. The streamlines in the interior show a slightly tilted ellipsoidal form. Flow direction is counter clockwise.

In the rest of the chapter this run is denoted as the reference case. Table 8.1 provides an overview on the solution type used for the reference case.

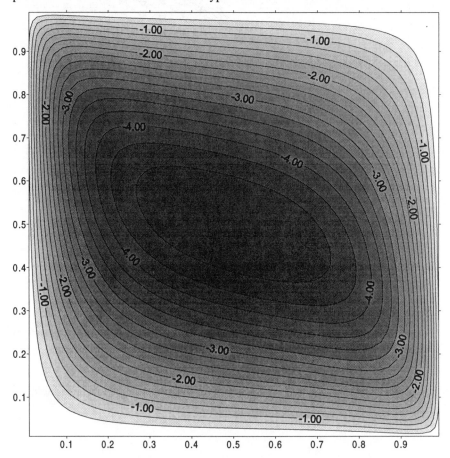

Fig. 8.5 : Streamfunction contours and filled areas for $H/L=1$ and $Ra=100$.

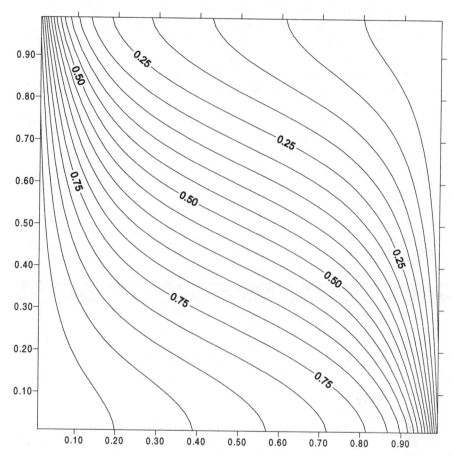

Fig. 8.6 : Temperature or salinity contours for $H/L=1$ and $Ra=100$

Parameter	Value/Type
Preconditioning factor flow solver	1.7
CG-type transport solver	BiCG
Preconditioning factor transport solver	1.5
Accuracy	10^{-4}
Solver precision mode	single
Number of inner iterations to reach accuracy	29

Table 8.1 : Solver type chosen for solving the reference case

The single precision solver BiCG produces results which are almost identical to those obtained with the double precision solver; the latter did not reach the required accuracy within 40 Picard iterations. A grid resolution analysis with a 100×100 mesh shows that there are differences around 1%.

The finite case is modeled with unit height and length $L \in \{1, 2, 4, 8, 16\}$. The grid for all values of L is 20×21 blocks, i.e. the grid-spacing Δx is varied in the input data.[1]

Results are compared by choosing $Ra \cdot \partial\theta/\partial x$ constant for all runs. The boundary condition is altered to account for the change of the length L; Ra could have changed instead as another alternative.

Table 8.2 compares results from numerical simulations with different finite values of L with the analytical solution of the system with infinite length. The velocity distribution in the unbounded system is given by equation (8.4). For all numerical models the horizontal velocity components in the center column (the 11th column) for the upper 12 vertical blocks. The absolute values of velocities in the remaining 9 blocks are identical to the values in the corresponding block in the upper half model - but the sign is opposite (negative).

z	L/H=1	L/H=2	L/H=4	L/H=8	L/H=16	eq. (8.4)
1/42	0.3143	0.4405	0.4736	0.4755	0.4757	0.4762
3/42	0.2705	0.3938	0.4261	0.4280	0.4282	0.4286
5/42	0.2303	0.3478	0.3787	0.3804	0.3806	0.3810
7/42	0.1936	0.3026	0.3313	0.3329	0.3330	0.3333
9/42	0.1599	0.2580	0.2839	0.2853	0.2854	0.2857
11/42	0.1290	0.2140	0.2365	0.2378	0.2379	0.2381
13/42	0.1004	0.1705	0.1892	0.1902	0.1903	0.1905
15/42	0.0737	0.1275	0.1419	0.1427	0.1427	0.1429
17/42	0.0483	0.0848	0.0946	0.0951	0.0952	0.0952
19/42	0.0239	0.0424	0.0473	0.0476	0.0476	0.0476
1/2	0.2E-6	0.2E-6	-0.5E-6	-0.6E-6	-0.7E-6	0.0
23/42	-0.0239	-0.0423	-0.0473	-0.0476	-0.0476	-0.0476

Table 8.2: Shallow layer: comparison of numerical results and analytical solution

The comparison makes clear that the numerical output is reasonable. For low values of L the solution is much different from the asymptotic analytical values. But the results converge with increasing length of the simulated system. The

---

[1] The input data for the model with Ra=400 can be found on CD-ROM in file *HorTrans.WTX*

system of maximum length (L/H=16) of the models, considered here, shows a difference between numerical and analytical results only in the fourth significant digit.

H/L	Ra	Nu^1	Nu^2	$\mid\Psi\mid_{max}$	Pe_x	Pe_z	N_{Iter}
10.	130.	1.030	1.030	1.6250	.0713	1.3083	3
10.	100.	1.019	1.019	1.2500	.0554	.9911	3
5.	130.	1.196	1.196	3.2285	.1605	1.4207	4
5.	100.	1.124	1.124	2.4883	.1250	1.0484	3
5.	80.	1.085	1.085	1.9941	.1011	.8189	3
1.	100.	3.150	3.211	4.7473	.6008	.9453	29
1.	80.	2.697	2.733	4.0558	.4815	.7278	17
1.	60.	2.234	2.252	3.2510	.3628	.5183	12
1.	40.	1.746	1.753	2.4009	.2430	.3198	
1.	20.	1.259	1.260	1.3662	.1250	.1426	3
1.	10.	1.079	1.080	0.7179	.0645	.0675	
1.	1.	1.000	1.000	0.0736	.0066	.0066	2
.5	100.	1.991	2.002	4.6869	.3312	1.1393	8
.5	80.	1.762	1.769	3.9020	.2686	.8818	6
.5	60.	1.510	1.513	3.0984	.2033	.6311	5
.5	40.	1.269	1.271	2.1696	.1368	.3962	3
.5	20.	1.085	1.086	1.1196	.0695	.1876	3
.1	100.	1.757	1.856	1.1977	1.0093	.0730	10
.1	70.	1.371	1.407	0.8489	.6830	.0463	5
.1	40.	1.124	1.133	0.4945	.3887	.2412	3
.1	20.	1.032	1.035	0.2495	.1957	.1152	3

Table 8.3 : Characteristics of systems heated and cooled from the side with varying aspect ratio H/L and Ra: Nu^1- Nusselt number 1^{rst} order approximation; Nu^2 - Nusselt number 2nd order approximation; $\mid\Psi\mid_{max}$- maximum of absolute values of streamfunction; Pe_x- maximum of grid-Péclet numbers in x-direction; Pe_z- maximum of grid-Péclet numbers in z-direction; N_{iter}-number of Picard iterations to solve for steady state

It can hardly be decided whether this deviance stems from the numerical method or is from the fact that the length in the numerical simulation is still far from $L=\infty$. This can be tested by further numerical experiments. Simulation runs for fixed L with refined grid-spacing give an indication on the accuracy of the numerical method.

Table 8.3 lists some post-processing parameters calculated from the FAST-C(2D) code for the purpose of a sensitivity analysis. The length is varied from $L=0.1$ to $L=10$, resulting in different L/H-ratios. As a second input parameter the Rayleigh number is varied. All numerical parameters were chosen identical with the reference case described above.

The following columns of the table provide an overview on the change of various crucial parameters. These characteristics are derived from the numerical model by post-processing. Nusselt numbers are calculated using 1st and 2nd order approximations (compare equations (4.43) and (4.44)). The maximum of the absolute value of the streamfunction is listed, as well as maxima of grid-Péclet numbers within the entire system (see equation (4.13)). The last column provides the number of Picard iterations needed to achieve the steady state with the required accuracy.

It can be noted in a first view that the number of iterations is low and thus the numerical effort quite marginal. If the modeler needs more reliable results, these can be produced with refined grid spacing and lower accuracy requirement in short execution times as well.

The maximum of the streamfunction and the grid-Péclet numbers also show higher values for higher Ra. The mean heat transfer, given by the dimensionless Nusselt number Nu for fixed H/L, increases with the Rayleigh number. The Nusselt numbers in the varied range of Ra are low for slender systems ($H/L=10.$) and for stretched systems ($H/L=0.1$). For all studied systems with same Ra heat transfer is highest for the system of equal length and height. Generally holds: if for a fixed value Ra systems with different H/L-ratio are considered there exists a system with maximum heat transfer.

It turned out to be difficult to obtain the system of maximum heat transfer with a coarse grid model using FAST-C(2D). For a fixed Ra number and fixed height H the length of the model L has been changed from one run to the other. For slender systems with small L and for extended systems with large L the steady state model provided results without problems. For systems with limited extension ($1.1<L<6.5$) the steady state model did not converge.

Starting from the final state in the former FAST-C(2D) run a transient simulation was started. The length was increased gradually and Nusselt numbers for the runs are denoted in Table 8.4. A maximum of heat transfer for the studied cases has been obtained for the system with length $L=2.25$.

A further increase of L leads to values for Nu which do not correspond with the results of steady state modeling - see deviations for $L=6.5$. Obviously the accuracy of $\varepsilon=10^{-4}$ is not restrictive enough. An accumulating error may also stem from the wide grid spacing.

L	Nu	s/t^2
0.375	1.60	s
0.5	2.00	s
0.75	2.68	s
0.9	3.00	s
1.0	3.17	s
1.1	3.34	s
1.2	3.42	t
1.3	3.66	t
1.5	3.84	t
1.75	3.98	t
2.0	4.04	t
2.25	4.05	t
2.5	4.05	t
2.75	3.99	t
3.0	3.96	t
3.5	3.81	t
4.0	3.68	t
4.5	3.55	t
5.0	3.44	t
5.5	3.35	t
6.0	3.27	t
6.5	3.20	t
6.5	2.32	s
8.0	2.08	s
10.0	1.77	s

Table 8.4: Model output for horizontal heat transfer Nu in systems with different length L

A further study on horizontal heat transfer with refined grids and accuracy is left to the reader.

---

[2] 's' for steady state, 't' for transient model run

9 Elder Experiment

9.1 Laboratory Experiment

J.W. Elder in his publication on *'Transient convection in a porous medium'* (1967b) describes two laboratory experiments. In the first experiment the rise of a hot blob is studied - this will not be discussed here. The second experiment is very similar to the constellation now well-known as Bénard convection; it has been treated in previous chapters. The classical convection analysis holds the upper and lower boundaries entirely at constant temperatures. Elder instead chooses a constellation, where the system is only partially heated from below.

In Elder's laboratory experiment half of the lower boundary is heated. The experiments have been performed in Hele-Shaw-cells. Hele-Shaw (1898) observed that two-dimensional flow of a viscous fluid in a narrow slot between two plates obeys a law similar to that described by Darcy (1856) (see chapter 3.1). If the distance between the plates is d, the corresponding permeability or conductivity for flow between the plates is given by the following formula

$$k = \frac{d^2}{12} \quad \text{and} \quad K = \frac{gd^2}{12v} = \frac{g\rho d^2}{12\mu} \tag{9.1}$$

The flow regime can be understood from the fact that friction becomes dominant relative to inertial forces when the gap between the plates becomes smaller - similar to the porous media situation. The use of glass or plexiglas as material for the plates allows isothermal lines to be observed by interferometry; the refractive index is temperature dependent. Streamlines can be visualized by using metal particles suspended in the fluid. Elder used photographic techniques to fix flow patterns at specified times. Hele-Shaw cells allow good visualization techniques. For that reason these cells are also used for research on low-viscous free fluids as well. Van Dyke (1982) comments: *'It is at first sight paradoxical that the best way of producing the unseparated pattern of plane potential flow past a bluff object, which would be spoiled by separation in a real fluid of even the slightest viscosity, is to go to the opposite extreme of creeping flow in a narrow gap, which is dominated by viscous forces'*

Most reported experiments note the distance d in the range of mm - in Elder's set-up: 4 mm. The total length of the cell used by Elder was 20 cm; the height was 5 cm and the model was electrically heated around the center of the bottomline across a distance of 10 cm. The gap between the two sheets of insulating material was filled with Silicon oil of type MS200 with a viscosity of 100 centiStokes.

9.2 Numerical Experiments

9.2.1 Elder's Model

Elder (1967b) himself presented the first numerical simulation of the laboratory set-up in the same publication that described the experiment. Elder refers to Wooding (1957) and uses the streamfunction formulation for the two-dimensional vertical cross-section, and a normalized temperature as transport variable. The following system is to be solved:

$$\omega = Ra \frac{\partial \theta}{\partial x}, \qquad \nabla^2 \Psi = \omega$$

$$F = \nabla^2 \theta - \frac{\partial \Psi}{\partial x} \frac{\partial \theta}{\partial z} + \frac{\partial \Psi}{\partial z} \frac{\partial \theta}{\partial x}, \qquad \frac{\partial \theta}{\partial t} = F \tag{9.2}$$

ω denotes the length of the vorticity vector (compare chapter 3.6). Note that the equations in the two rows can be combined. This eliminates the vorticity ω and F and yields the formulation derived in chapter 3. Fig. 9.1 shows the boundary conditions to be applied. For streamfunction Ψ a Dirichlet-condition is prescribed on all edges. Concerning heat transport, the vertical edges obey the Neumann condition $\partial \theta / \partial x = 0$. The physical meaning is that there is no diffusive heat flux across these boundaries: they are perfectly insulating because the flow boundary condition allows no additional advective flux either.

Normalized temperature has the Dirichlet-condition $\theta = 1$ in the heated part at the lower edge of the cell and $\theta = 0$ at the upper edge of the cell. On the not heated part of the lower boundary there is most reasonably a Neumann no-flux condition $\partial \theta / \partial z = 0$. It should be noted that Elder's original publication proposes different conditions: $\theta = 0$ everywhere outside the heated region, i.e. at the top and vertical boundaries, too. There has been a broad discussion within the HYDROCOIN project about the boundary conditions. Holzbecher (1986) remarks that either $\theta = 0$ or $\partial \theta / \partial n = 0$ at all unheated boundaries are both unrealistic.

The transient simulation treats the case where the heater on the bottom boundary is switched on at time $t=0$. The initial condition is the constant function $\theta\equiv0$. Final simulation time is $t_{max}=0.1$, when small changes in the numerical solution indicate that steady state is almost achieved.

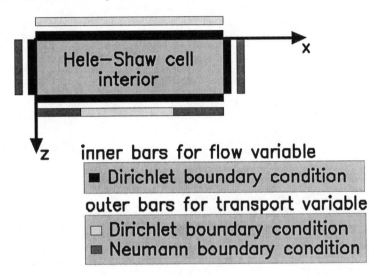

Fig. 9.1: Boundary conditions for the Elder-experiment in streamfunction formulation; transport variable is normalized temperature; $\Psi=0$ on all edges

Elder treats the finite difference representations of the four equations (9.2) in the given order. Explicit formulae result from the 1st, 3rd and 4th equation. The numerical algorithm resulting from the latter two equations is identical to the explicit method or forward-in-time discretization from chapter 4.

The discretization of the second (Poisson-) equation remains to be solved by a more sophisticated method. Elder in the original paper uses Leibmann's extrapolated method connected with an alternating direction of scan. The algorithm, an extended version of the overrelaxation method, is described in more detail in an earlier paper by Elder (1966). He reports that eight iterations are sufficient to achieve a precision of 1%. The Courant- and Neumann-criteria as restrictive conditions for the timestep are mentioned. The Courant condition, with reference to Courant e.a. (1928) often called Courant-Friedrichs-Lewy condition, is found to be an *'excellent guide'* in choosing Δt.

The original publication gives contour plots for Ψ and θ for six various times after the sudden switch on of the heater. The problem is re-calculated with a finer grid and smaller timesteps using FAST-C(2D). The results for the half system, as shown in Fig. 9.2, agree qualitatively. More details about the FAST-C(2D) model are given in chapter 9.2.2.

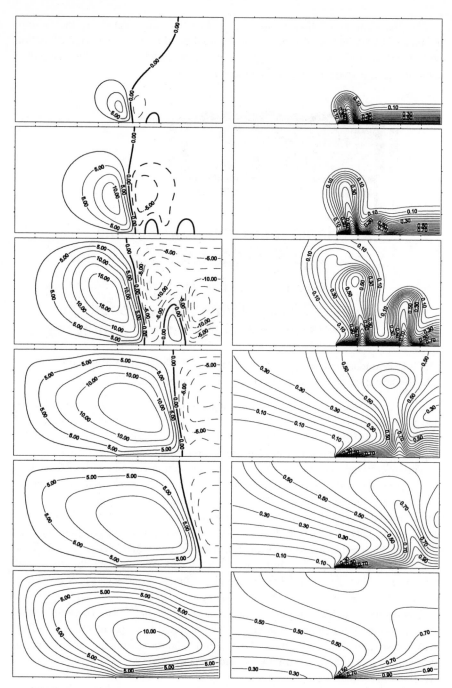

Fig. 9.2: Elder problem; streamfunction contours, isotherms for six timelevels
(t=.005, .01, .02, .05, .075, .1); calculated using FAST-C(2D) on a coarse grid

At first (t=0.005 and t=0.01) two small eddies emerge on both sides above the heater ends initiated by rising fluid. A complicated pattern with two large and two small eddies in both sides can be observed at time t=0.02. At time t=0.05 the two small eddies become almost invisible, and finally (t=0.1) the two big eddies melt together to one roll. The steady state in the whole cell is thus characterized by two rolls. The left circulates counter-clockwise, the right roll circulates clockwise. The center of the system has upward flow.

Despite the symmetry of the problem with respect to the center vertical axis, the calculated results show a slight asymmetry increasing with time. The reason for this could probably be found in the insufficient end criterion in the solution of the Poisson equation. If the accuracy is only 1% (see above), the propagation and increase of the errors in the transient simulation is likely to cause deviations as observed. Nevertheless the asymmetry does not change the qualitative description given above.

Elder's post-processing calculates vertical heat flux and Nusselt numbers for the reference case with Rayleigh number Ra=400. Transient change of the maximum value of streamfunction is presented for $Ra \in \{100, 200, 400\}$.

9.2.2 FAST-C(2D) Model

The FAST-C(2D) code is based on a discretization of one flow and one transport equation (see chapter 3.7 and chapter 4). It can be derived from equations (9.2) by eliminating ω and F. As mentioned above: the selection of the explicit option results in an algorithm very similar to the one used by Elder. There are differences in the solution of the Poisson-equation. FAST allows to choose from various conjugate gradient solvers which treat sparse linear systems resulting from the discretization directly. Differences in the discretization may arise from different discretization schemes, what mainly concerns the first order derivatives. Elder does not indicate which of the most common schemes - upwind or central - he applied for his calculations.

First results obtained from a former version of FAST-C(2D) have been published in Holzbecher (1991). The model used grids of 40×25 and 60×25 blocks then. New model runs have been performed using different grids, and are presented in this chapter[1]. Only half of the system is modeled. The boundary condition at the vertical axis through the center of the Hele-Shaw cell is of no-flow type both for fluid and heat flow. In the analytical formulation the conditions Ψ=0 and $\partial\theta/\partial x$=0 are required as boundary conditions. At first the grid-spacing from Elder's model was adopted, i.e. Δx=Δz=0.05 in the dimensionless formulation. Elder (1967b) used a 80×20 regular rectangular mesh for the entire model region and regular timesteps Δt=0.0025.

---

[1] The input data for the model can be found on CD-ROM in the file *Elder.WTX*

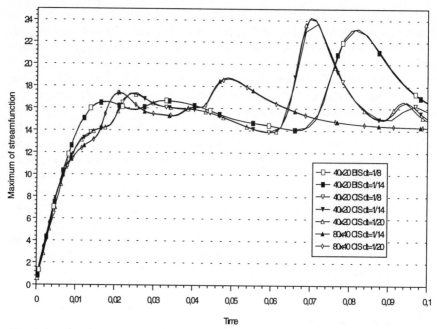

Fig. 9.3: Temporal development of streamfunction maxima for grid spacing 0.05 (40×20 grid in half cell)

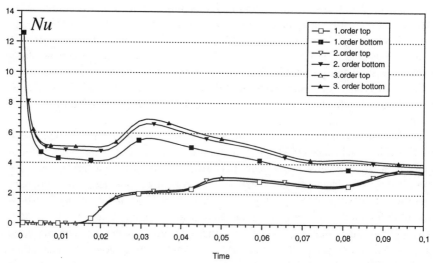

Fig. 9.4: Temporal development of Nusselt numbers for grid spacing 0.05 (40×20 grid in half cell)

The timesteps used in the FAST-C(2D) model are increased during the simulation run. Effects of grid refinement and discretization techniques (BIS and CIS - see chapter 4) are tested. The presented figures show the transient development of some system variables for the Ra=400. The same system has been studied by Elder in his original publication. The deviance between Elder's results and FAST-C(2D) output is smaller for smaller values of Ra.

Fig. 9.3 shows the temporal development of streamfunction maxima for different grids, discretization methods and timesteps. Obviously the model is almost insensitive to the variation of the timesteps in the range chosen for the simulation (the value in the notation 'dt=' given in the figure can be used only relatively). In contrast the discretization method (BIS or CIS) and the grid spacing are important. Significant differences can be observed for BIS and CIS method on the coarse grid (40×20 blocks) and for the CIS-method applied on different grids.

Fig. 9.4 shows the temporal development of the Nusselt number at top and bottom boundary of the system using different approximation methods (see chapter 4). Nu representing heat transfer is high immediately after the start of the experiment, while it takes a while until the heat transfer becomes visible at the top boundary. Differences between the approximations can clearly be recognized for the heat transfer at the lower boundary, while they are marginal for the top of the model. 1st order approximations give smallest values, 2nd order values are highest and 3rd order values lie in between. At the end of the simulated time period 1st order Nu-approximations at the lower and upper boundary are almost identical. Obviously the steady state, where heat transfer is the same at both edges are equal, is nearly but not yet reached.

A grid resolution study has been made using three different grids for the half system: 40×20, 80×40 and 160×80. The following figures show the dependence of the results on grid resolution, for the BIS method and the CIS discretization method.

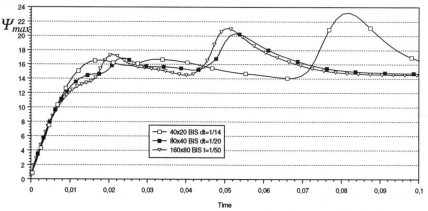

Fig. 9.5 : Transient FAST-C(2D) results using BIS for different grid resolution

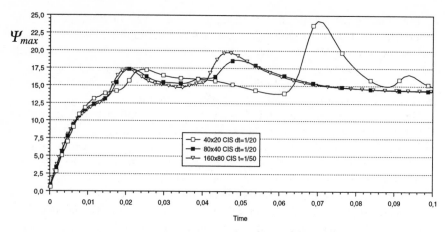

Fig. 9.6 : Transient FAST-C(2D) results using CIS for different grid resolution

That results vary with grid discretization method, has already been mentioned as comment on Fig. 9.3. But it turns out for both of the studied methods that the results for the finer grids are smoother than the output for the coarse grid. Results for the fine grids are almost identical at the beginning of the simulation and during the final approach of the steady state - that holds for comparison of results from CIS and BIS runs respectively. And it holds as well, when output of the two finest grids are compared. There are some deviations in the intermediate time range still. It can be concluded that the solution for the finest mesh is almost grid independent. Marginal changes on streamfunction maxima, when the final time level is approached, indicate that the system is very likely near its steady state.

In order to confirm these statements two additional runs have been made with an additional refinement. A grid of 160×80 equidistant blocks was selected and run with different timesteps. The outcome from both discretization methods are gathered in Fig. 9.7. For a reduced but coarse timestep there appear again fluctuations, which are smoothed out, when a finer timestep is used. Most similar are here results from BIS with wide Δt and CIS with small Δt.

The comparison of the temporal development of streamfunction maxima give an indication only the flow pattern. Thus rise and fall of the Ψ-curves at certain times can be related to upcoming or vanishing eddies in the solution. This is not to be done here in detail. Of interest is that the fluctuating curve in Fig. 9.7 (CIS t=1/26) represents a different final convection pattern than the other three runs. While there is one eddy remaining in the half system in the first case, there are two eddies in the almost steady state in the other cases (see also Table 9.1).

Despite some differences in the intermediate period both discretization methods deliver the identical steady state solutions. These show two eddies in the half cell, i.e. four eddies in the entire cell.

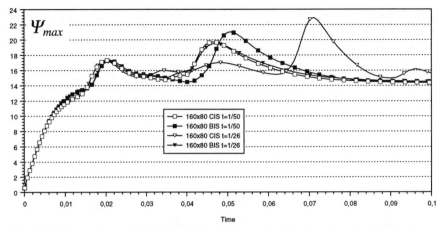

Fig. 9.7 : Comparison of FAST-C(2D) results on the finest grid, using BIS and
CIS discretization methods and different size of timesteps

Table 9.1 provides an overview on the number of rolls in the final state for the
simulations shown in the figures. Obviously grid refinement leads to solutions
which show two eddies in the half system at the final state, i.e. four eddies in the
entire Hele-Shaw set-up. There is downward flow along the vertical symmetry axis
of the model in the final state.

Grid	Discretization	Max. Timesteps	No. of eddies
40×20	CIS	1/20	1
,,	BIS	,,	1
80×40	CIS	,,	1
,,	BIS	,,	2
160×80	CIS	1/26	1
,,	CIS	1/50	2
,,	BIS	1/26	2
,,	BIS	1/50	2

Table 9.1 : Number of eddies in the half system of the Elder-experiment at $t=0.1$
simulated with FAST-C(2D) using different grid spacing and discretization options

The obvious sensitivity on grid resolution in the studied range of spacings
allows the conclusion that the results published by Elder (1967b) are *not* grid
independent (Elder used a spacing equal to the one in the 40×20 grid here). The
flow patterns given in the original publication need to be revised as well. The main
difference is that the final state for simulated time $t_{max}=0.1$ has still two eddies in
the half system. This pattern has been calculated using FAST-C(2D) with a fine

grid (160×80) and small timesteps. Elder's solution in contrast has two eddies in the entire model, i.e. just one in the half system.

The comparison with the figure in Elder's publication shows curves to be similar in many respects (Fig. 9.8). The same spatial and temporal discretizations have been used. i.e. a 40×20 grid for the half-cell and Δt=0.0025. The outlook of the graph is qualitatively identical. But there are differences concerning the time and amount of change increase with simulated time.

Probably Elder with his calculations would have obtained similar solutions as presented above, if he would have made a grid refinement study. Restricted capacities of computers at that time prevented him from doing that.

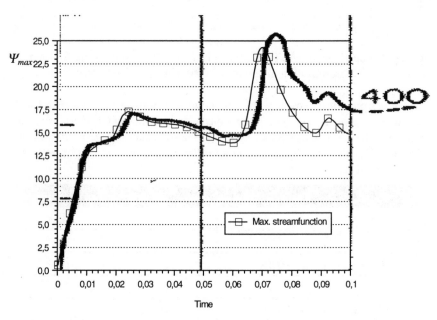

Fig. 9.8: Comparison of the FAST-C(2D) results with original publication by Elder (1967b) with identical space- and time-discretizations

9.2.3 Further Models

Modeling before HYDROCOIN

Diersch (1981) treated the saline convection case using a *'primitive variable finite element'* formulation with pressure, velocities and concentration as unknown variables. Only half of the model region was simulated, which is justified because of the symmetry of the problem. The region is covered by 63 eight node elements of Serendipity type. Diersch remarks that he tried coarser meshes without success: the starting phase with small eddies could not be reproduced. Reviewed from grids used in numerical models nowadays, the 63 element structure looks much too coarse. The coarse mesh is probably the reason, why the resulting concentration contours show less details than those discussed in the HYDROCOIN project.

Modeling in HYDROCOIN

The Elder experiment was selected as a test case for model validation in the international HYDROCOIN project (HYDROCOIN 1990). The participating teams in the project were requested to use either the dimensionless parameters from the original Elder paper (1967b) or to model an equivalent situation of saline convection (Heredia/Holzbecher 1986). The convective motion can be initiated by a salt gradient instead of a temperature gradient, when the salt is dissolved on the top boundary. The lower boundary should have salt removed from the system; this is hardly realizable in a laboratory experiment. The analogy requires the additional assumption that direction-independent diffusion governs the mass flux from concentration gradients as described in classical Fick's law. Field and laboratory experiments indicate that direction-dependent dispersion mostly governs this type of processes. The conceptional model as described makes sense for testing numerical codes. Parameters for the saline convection case as specified in HYDROCOIN level 2 case 2 are given in the following table.

Symbol	Parameter	Value	Unit
k	permeability	$4.845 \ 10^{-13}$	m^2
φ	porosity	0.1	-
μ	viscosity	$1.0 \ 10^{-3}$	Pa s
g	gravity	9.807	m/s^2
ρ_{fresh}	fresh water density	1000	kg/m^3
$\Delta\rho$	max. density difference	200	kg/m^3
D	diffusivity	$3.565 \ 10^{-7}$	m^2/s
H	height	150	m

Table 9.2: HYDROCOIN test case parameters for the saline convection case

Times 0.005, 0.01, 0.02, 0.05, 0.075 and 0.1 of the dimensionless formulation correspond with times 1, 2, 4, 10, 15 and 20 years. It is convenient to compare the number and the shape of eddies. This can be done for results of streamfunction only, but Ψ as unknown variable can usually not be selected.

Thus contours of the transport variables temperature or salinity are taken for comparison in most publications. Results from four different codes, including FAST, have been submitted to the HYDROCOIN secretariat: from the FD-code

SWIFT in the version of DWDM (1976), and from the FE-codes CFEST (1982) and METROPOL. CFEST-results are reported additionally by Roome (1989), and METROPOL-results by Leijnse/Hassanizadeh (1989). The final HYDROCOIN report makes a visual comparison of contours and concludes: *'although the individual isopleths show some variations, the agreement between the different results is judged to be satisfactory'*.

Although the general features of the Elder experiment - the emergence of eddies starting from the heater ends and their development towards a two roll steady state - are reproduced by all simulations, there are great differences in details of the convective flow pattern. There are differences between results obtained with different codes - or even with the same code using different options (see the previous chapter). Moreover there are deviations from the results reported by Elder in his original publication.

Several reasons can be found for these differences. FAST-C(2D) code and Elder's simulations require the validity of the Oberbeck-Boussinesq assumption. Other codes are based on a more general set of equations. Still differing distributions for θ are obtained with codes not based on the Oberbeck-Boussinesq assumption. Different discretization approaches (FD or FE in rectangular grids, FE in triangular grids) may cause different results. The dependence on grid refinement has seldom been studied: thus codes based on the same discretization method of the same equations may produce different results, if the grid is not identical. Accuracy criteria may also play a role.

The reason for the differences between Elder's results and those from the HYDROCOIN participants cannot be found in the fact that the analogy between heat and mass transport is not given in reality. The underlying fundamental laws are assumed to be identical in the case formulation. The question, if the modeling of the Elder-experiment is satisfactory depends on the requirements. If it is necessary to get the details of the convective motions and their temporal change, the current state of the modeling is not yet sufficient. But the exact correspondence between results obtained with different codes was not a goal of the HYDROCOIN project.

All what is mentioned above is of concern for the verification of codes. There may be further difficulties, when numerical results are compared with those from experiments. Elder himself skipped this important point, which would be the first step to validation. But to do that properly the experiment would have to be re-run. The original Elder experiment is not sufficiently documented to give a test case for validating models for density driven flow - a similar statement concerning HYDROCOIN test cases is made by Larsson (1992).

Modeling since HYDROCOIN

Voss/Souza (1987) in their FE discretization cover the entire model with 1100 elements and 1170 nodes. The SUTRA-code for solute transport problems has concentration and pressure as unknown variables and uses bilinear basis functions.

Voss/Souza in fact take the input data from the HYDROCOIN test case definition (Heredia/Holzbecher 1984). But their model formulation requires flow boundary conditions to be defined in terms of pressure. There are Dirichlet conditions $p=0$ specified in both corners on the upper edge. The initial no-flow condition is given by the hydrostatic pressure distribution. The comparison of isohalines from the SUTRA model with Elder's isotherms can also be found in the book of Ségol (1994).

Holzbecher (1991a) compares results from four models. SWIFT and FAST are applied with a 1000 block grid, SUTRA with 1100 elements. The 20% and 60% contours for $t=0.02$ and $t=0.2$, compared with those given in the Elder's original publication (1967b), show some differences.

A better impression of the deviations is achieved if the *'fingers'* in some isopleths are counted. If the 0.2-contour is taken, for time $t=0.02$ there are five differently developed fingers in the original Elder results. The numerical simulations with different codes show different numbers: SWIFT (5), FAST (3), CFEST (5), METROPOL (4), SUTRA (5), FEFLOW (4). At time $t=0.05$ there are three fingers in the Elder results and in other simulations: SWIFT (3), FAST (3), CFEST (5), METROPOL (4), SUTRA (2), FEFLOW (4).

Kolditz (1994) takes the Elder experiment as a benchmark for the variable density FE code FEFLOW. His extensive work studies the sensitivity of the results to the selection of discretization schemes, to grid spacing, to the Oberbeck-Boussinesq assumption, to boundary conditions, to timestepping schemes and to solvers of linear equations. Only some main points are repeated here. Kolditz provides the thermal convection in the original experiment with a set of data with physical dimensions shown here in Table 9.3.

Symbol	Parameter	Value	Unit
d	channel aperture	0.004	m
L	length	0.2	m
H	height	0.05	m
k	permeability	$1.333 \ 10^{-6}$	m^2
β_T	thermal expansion coeff.	$6.372 \ 10^{-6}$	1/K
D	heat diffusivity	$1.041667 \ 10^{-7}$	m^2/s
v	kinematic viscosity	10^{-4}	m^2/s
Ra	Rayleigh number	400	-

Table 9.3: Parameters for thermal convection in the Elder-experiment

At first Kolditz (1994) presents FEFLOW simulations with grids previously used by Elder (1967b) and Voss/Souza (1987) and which contain 1700 and 1100 elements respectively. It turns out that output for these coarse meshes is not independent from the grid parameters. Further runs with FEFLOW with finer grids show that results with 4400 element grid and a 9900 element grid are nearly identical. This coincides very well with the statement of Leijnse (1995) about some recent calculations with METROPOL. He stated that grid-independent output can be obtained for grids with more than 4000 elements in the half-cell.

As further result Kolditz reports that the use of the upwind-scheme produces smoother contours of isohalines. Local 'fingers' are smoothed out and the further development of these disturbances is not possible. The details of the transient patterns are quite different and the final steady state looks quite different, too. Even small differences, produced using different solvers for the linear equations, accumulate to visible differences in the shape of the final contours.

In a second numerical analysis Kolditz (1994) demonstrates that the deviations from the choice of numerical methods is smaller on the finest grid. This result is expected because numerical results with the refinement of grid spacing converge towards analytical solutions. There may be divergence in exceptional cases.

Different time-stepping procedures are studied by Kolditz (1994) (see chapter 4.2):

- fully implicit ($\kappa=0$)
- Crank-Nicolson ($\kappa=0.5$)
- predictor-corrector schemes

Higher order schemes, like Crank-Nicolson or predictor-corrector, produce nearly identical results. Slightly different output is obtained for the lower order implicit scheme. The predictor-corrector method, as applied by Kolditz (1994) using FEFLOW, did not converge in all cases.

The final state obtained by different methods is almost identical, even if models are compared for which the Oberbeck-Boussinesq assumption is applied or not. The final flow system has two rolls in the half system and four rolls in the entire Hele-Shaw cell. There is upward flow along the vertical axis of symmetry. The main characteristics of the flow pattern coincide well with those obtained by the FAST-C(2D) model in chapter 9.2.2. Note that in FAST-C(2D) thermal convection is simulated. In contrast Kolditz (1994) tackled the saline case.

Oldenburg/Pruess (1995) take the Elder experiment for the verification of TOUGH2 code - the saline convection case is modeled using the input data defined in the HYDROCOIN project (Table 9.2; Heredia/Holzbecher 1984). It is reported that TOUGH2 results for a similar grid agree with those from Elder (1967b) and Voss/Souza (1987). It has been verified that output for the half system is identical to that of the entire system using the same grid spacing.

A half cell model was used for a grid resolution study is performed by Oldenburg/Pruess (1995). A uniform grid with 60×32 (=1920) and a graded grid with 84×42 (=3528) blocks yield the same solutions. This is checked by a visual comparison of flow field and salinity contours after 2, 10 and 20 years simulated time. The final state (20 years simulated time) shows downward flow in the cell center. This contradicts the results obtained with TOUGH2 and other codes for the coarse grid as mentioned above. Indeed the shown flow pattern after 20 years is characterized by two eddies in the half-cell, while the low-resolution figures show a single roll on both sides of the symmetry axis. Oldenburg/Pruess (1995) state that the high-resolution results agree with the original Elder experiment. This is not true, for Elder obtained upward flow in the center for thermal convection - while here the saline convection case is treated and the situation should be vice versa. Solutions obtained by using FAST-C(2D) code for high resolution grids (see chapter 9.2.2, Table 9.1) show a two-eddy pattern in the half-system at time $t=0.1$, which is the final time chosen by Elder. If the simulation continues, the inner eddy becomes weaker and finally disappears.

It seems to be important to compare grid-independent results obtained with different codes or different options from the same code. If only model outputs for different coarse grids are compared, it cannot be distinguished if deviations stem from different grids or from the different algorithms.

The intercomparison of results from different algorithms is a important task. The study of the origins of differences in model outputs is a time-consuming task; it could only partially be done in the HYDROCOIN project. It is an important future goal to show the reliability of modeling software.

9.3 Related Problems

Further experimental and numerical work with a system very similar to the Elder experiment has been presented by Horne/O'Sullivan (1974). Unlike Elder, only half of the symmetric system has been treated, both experimentally (extensions: 30×20 cm) and numerically. The crucial difference is that the unheated part of the lower surface has been cooled to the same temperature as on the upper boundary. Fig. 9.9 gives a graphical sketch of the boundary condition. The heated part has been changed from ¼ to ½, to ¾ and to the entire length of the bottom line. Horne/O'Sullivan varied the Rayleigh number up to a value of 1600. The main interest of the publication was to study oscillatory convection.

The numerical calculations of Horne/O'Sullivan (1974) use a finite-difference representation of the system

$$\nabla^2 \Psi = -\frac{\partial \theta}{\partial x} \qquad \frac{\partial \theta}{\partial t} = \nabla^2 \theta + Ra\left(-\frac{\partial \Psi}{\partial x}\frac{\partial \theta}{\partial z} + \frac{\partial \Psi}{\partial z}\frac{\partial \theta}{\partial x}\right) \qquad (9.3)$$

which can be obtained from the system used by Elder and FAST-C(2D) by the transformation $\Psi \rightarrow Ra \cdot \Psi$. The discretization of 1st order derivatives uses higher order Arakawa schemes: the 9-point-discretization in the lower- and uppermost lines of the grid and the 13-point-discretization at other grid points. The node-centered grids are rather coarse: 17×17 and 33×33 blocks.

Table 9.4 provides results from the original paper. For various Rayleigh numbers and heating regimes the characteristic of convective flow is shown. For most cases with rising *Ra* a switch from steady to fluctuating or oscillatory patterns can be observed. Note that the situation in the second column represents the Bénard case (see chapter 5). The result coincides roughly with later findings that there is a Hopf bifurcation at *Ra*≈380 (see chapter 7).

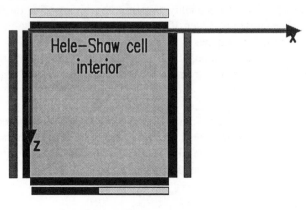

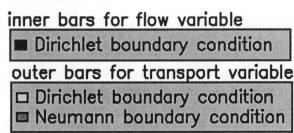

Fig. 9.9: Sketch of boundary conditions for the Horne/O'Sullivan experiments

Ra	All heated	¾ heated	½ heated	¼ heated
250	steady		steady	steady
375	fluctuating		steady	steady
500	fluctuating	steady	oscillatory	oscillatory
750	fluctuating	steady	oscillatory	oscillatory
1000	fluctuating	steady	oscillatory	oscillatory
1250	fluctuating	steady	oscillatory	oscillatory

Table 9.4: Flow regimes for systems (entirely and partially) heated from below (according to Horne/O'Sullivan 1974)

The constellation nearest to the Elder experiment (heating half of the bottom and Ra=375) yields a steady state, while the next to nearest (heating half of the bottom and Ra=500) shows an oscillatory regime. The marginal case is determined for Ra=480.

The cases described by Horne/O'Sullivan can easily be modeled using the FAST-C(2D) code. Using finer grids and timesteps Table 9.4 probably will have to be revised at some places. Even the fine grid of the cited authors is rather coarse viewed from nowadays computer capabilities. A sensitivity analysis concerning grid spacing and time-stepping I leave to some interested reader.

10 Geothermal Flow (Yusa's Example)

10.1 Hypothetical Situation and Analytical Description

An idealized two-dimensional situation of geothermal flow has been proposed by Yusa (1983). The characterization of flow patterns in the hypothetical cross-section can be used for code verification. The model region is bounded from below by an impermeable layer and at the vertical sides by water divides. The flow regime on the top is predetermined by a hydraulic gradient which is varied in a sensitivity study from 0 to 5%. The intention of Yusa is to consider the influence of potential flow on one side and thermal convective flow on the other side.

For a constant density fluid the flow field in such a trough situation is well known from an analytical solution. According to Tóth (1962) hydraulic head is given by:

$$h(x,z) = h_{min} +$$

$$+(h_{max} - h_{min})\pi \left[\frac{1}{2} - \frac{4}{\pi^2} \sum_{i=0}^{\infty} \frac{\cos[(2i+1)\pi x / L]\cosh[(2i+1)\pi(H-z)/L]}{(2i+1)^2 \cosh[(2i+1)\pi H / L]} \right] \quad (10.1)$$

Yusa (1983) studies the influence of geothermics by introducing a heat source at the bottom of the system. Across a horizontal distance of length L, the underlying impermeable layer is kept at a high temperature due to an increased geothermal flow from the earth interior. The heated part lies on one side of the system adjacent to the water divide with higher hydraulic head.

Yusa's analytical model formulation uses a dimensionless streamfunction and normalized temperature. His definitions are mostly identical to those introduced in chapter 3. There is only one major difference; using the transformations

$$t \to \frac{H^2}{D} t \qquad x \to \frac{x}{H} \qquad z \to \frac{z}{H} \qquad T \to \theta = \frac{T-T_0}{\Delta T} \qquad (10.2)$$

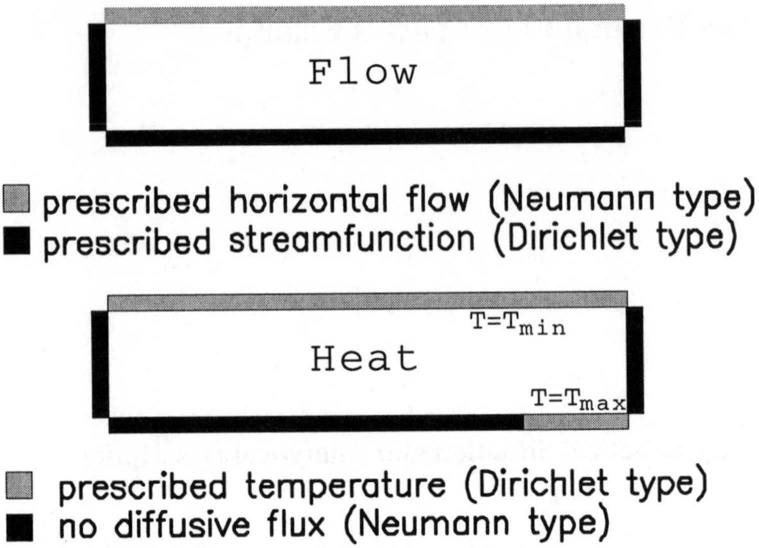

Fig. 10.1: Boundary conditions in (Ψ, θ)-formulation for the Yusa example

Yusa obtains the system of differential equations:

$$\frac{\partial^2 \Psi}{\partial x^2} + \frac{\partial^2 \Psi}{\partial z^2} = R\hat{a}\frac{\partial \theta}{\partial x} \qquad \text{with } R\hat{a} = \frac{k\Delta\rho g H}{D\mu}$$

$$\frac{\partial \theta}{\partial t} = -\gamma v \nabla \theta + \nabla^2 \theta$$

(10.3)

γ is taken neither into the transformation (3.40) nor into the Rayleigh number definition. The system (10.3) is equivalent to the equations derived in chapter 3 (compare with (3.41) and (3.45)). In order to be consistent in the following the Rayleigh number definition from chapter 3 is used.

Fig. 10.1 provides a visual representation of the boundary conditions for the formulation using streamfunction and normalized temperature as unknown variables. Fig. 10.1 is valid for both systems: the system used by Yusa (1983) in his earlier publication, and the system on which the FAST-C(2D) code is based. The streamfunction formulation is used for the description of flow. All boundary conditions with prescribed streamfunction are specified with the same value (Ψ=0). The entire boundary line - from the top left corner of the model down, along the bottom, up to the top right corner - becomes a streamline. It is a limiting streamline, across which there is no fluid flow.

The mathematical point of view notices an inconsistency at both corners of the top model line when a non-zero value is required for the prescribed horizontal flux along the line. However the argument is not relevant because analytical and numerical solutions converge and deliver physically reasonable solutions. Formula (10.1) does not converge even in closed intervals (Holzbecher 1991a) and the derivative is not a continuous function - the equivalence of both these properties can be shown. Some remarks on the very same point have been made within the discussion of the salt-dome problem in the HYDROCOIN project (see chapter 13).

Symbol	Parameter	Value
T_{min}	minimum temperature	20°C
T_{max}	maximum temperature	250°C
H	height	1000 m
L	length	5000 m
$\Delta\rho$	density difference	230 kg/m^3
μ	dynamic viscosity	2. 10^{-4}kg/m/s^2
k	permeability	10^{-14} m^2
D	thermal diffusivity	10^{-6} m^2/s
γ	ratio of heat capacities	2
Ra	Rayleigh number	113
$\partial h/\partial x|_{top}$	hydraulic gradient	0, 1%, 3%, 5%

Table 10.1: Parameter for the Yusa geothermal flow example

Table 10.1 provides an overview on parameters describing the system on geometry and temperature range. Homogeneity is assumed for all parameters in the example.

The transient model starts with a hypothetical situation of constant temperature and streamfunction fields. Strictly speaking. the zero flow initial condition is inconsistent with the top boundary conditions if there is a non-zero hydraulic gradient. The flow patterns, that appear for certain parameter constellations - Yusa's main concern - are not influenced by the initial condition.

10.2 Flow Pattern Characterization

Four different cases are studied where the prescribed potential gradient I is changed. In the no-gradient case holds $I=0$. In further studies I takes the values 0.01, 0.03 and 0.05. The flow patterns and corresponding isotherms depend strongly on the gradient I. The final timelevel shows a near steady state in three of the four cases.

Further simulations of the test-case have been reported by Yano (1989), Holzbecher/Yusa (1995) and Springer (1995). Yusa (1993) and Holzbecher/Yusa (1995) use the finite difference method for the streamfunction formulation, Yano (1989) and Springer (1995) apply the finite element method on the pressure formulation.

Yano (1989) calculated the Yusa example using a triangular mesh. In contrast to the above mentioned codes, a pressure/enthalpy formulation is used for modeling. The transport equation is formulated as:

$$\left((1-\varphi)\rho_r \frac{C_r}{C} + \varphi\rho \right) \frac{\partial h_e}{\partial t} = \mathbf{v} \cdot \nabla h_e - D\nabla^2 h_e \qquad (10.4)$$

Some further parameter values had to be defined due to the different formulation of the analytical model. Table 10.2 gives a list of parameters and their values used by Yano (1989), in addition to the original definition of the example.

Symbol	Parameter	Value
φ	Porosity	0.1
C_r	Spec. heat capacity (rock)	0.152 cal/g/°K
λ	Thermal conductivity	$5 \cdot 10^{-3}$ cal/s/cm/°K
ρ_r	Rock density	2700 kg/m^3

Table 10.2: Additional parameters as specified by Yano (1989)

Springer (1995) used the THERMOD code, written by himself. He took a regular grid with 80×20 elements for a system of 4000 m length. As in some calculations of Holzbecher/Yusa (1995), the system was shortened on the left side by 1000 m - a region mostly irrelevant with respect to the aim of studying the interaction of potential and convective flow. The following presents Springer's results and comments on differences from other models.

An ascending flow region emerges in the free convection case, where no head gradient is set on the top boundary. If temperatures are plotted for the same depth level, highest values can be found in the ascending flow region. When the simulation proceeds the ascending flow region moves horizontally towards the adjacent closed boundary, until only one half convection cell remains (Fig. 10.2). Upward flow of hot water can be observed in a narrow region in the vicinity of the vertical boundary. Velocities remain very small in the far field - in the figure near the left boundary.

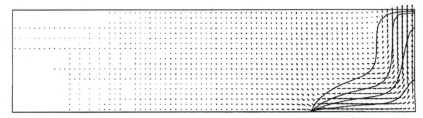

Fig. 10.2: Steady state for the reference case without natural hydraulic gradient[1]

---

[1] Figures 10.2-10.8 by courtesy of Jörn Springer, GeoForschungsZentrum (GFZ) Potsdam, Germany

Ascending flow can be observed in the beginning of all simulations. The details of the hydrodynamics are very different. Fig. 10.3 visualizes the phenomenon clearly for the $I=0.01$ case. The upward flow of hot water is mixed with the natural flow regime - in the figures from right to left.

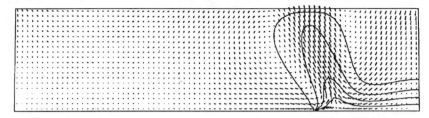

Fig. 10.3: Flow pattern for 1% hydraulic gradient after the start of the simulation

Different flow pattern appear with increasing simulation time. The $I=0.01$ case shows the emergence of a second ascending flow region near the right boundary (Fig. 10.4). A region with downward fluid flow is established in between. The observation corresponds that the temperature distribution shows lower values at same depths. In the entire region two convective eddies are established as effect of the thermal source at the bottom of the cross-section. Another half-eddy, induced by the potential gradient, is still visible on the left hand side of the system.

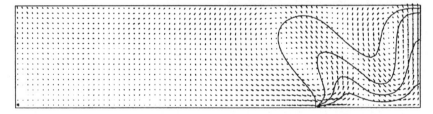

Fig. 10.4: Intermediate flow pattern for 1% hydraulic gradient

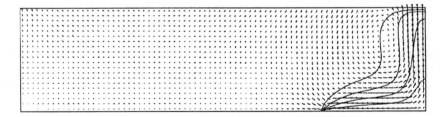

Fig. 10.5: Steady state for the low gradient case, as calculated by Springer (1995)

In the case of $I=0.03$, the convective flow ascends with an increased angle to the positive horizontal axis (Fig. 10.6). This phenomenon is a clear effect of the greater influence of potential flow. Where the top boundary is approached by ascending flow from below, temperatures are lower than in the cases described previously.

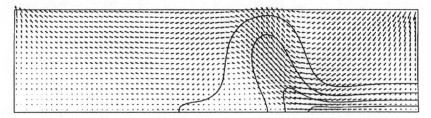

Fig. 10.6: Flow pattern for 3% hydraulic gradient after the start of the simulation

For the I=0.03 case no model reaches a steady state, corresponding to Springer's results shown here. There are transients that move upward and away from the heated part of the boundary (Fig. 10.7). The hot plumes become cooler and finally vanish. The details of the transient motions are quite different and incomparable.

There are some differences in the basic flow pattern, too. Yusa (1983) and Holzbecher/Yusa (1995) report on the emergence of an ascending flow region in the vicinity of the right hand side boundary. This is a similarity with cases previously described, but could not be confirmed by Springer (1995). He got a downflow regime at the same boundary (Fig. 10.7).

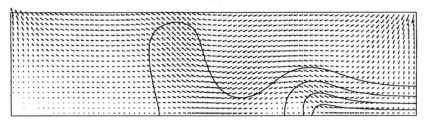

Fig. 10.7: Final state for medium gradient case, as calculated by Springer (1995)

If a 5% hydraulic gradient exists at the top, the potential flow regime dominates. Hot water plumes apparently cannot ascend against the strong potential flow that is mainly horizontal. Hot fluid isotherms can only be observed near the bottom boundary. For long simulation times the temperature increases on the left boundary. Hot water flow is upward due to the no-flow condition at the vertical boundary.

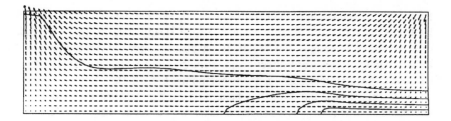

Fig. 10.8: Steady state for strong gradient case

The comparison of the results of Holzbecher/Yusa (1995) with the output from Yusa (1983) shows that there are some differences concerning the transient development. These are reasonable, if the differences of the two models are considered. The earlier calculations, made on a mainframe computer, used a fine grid with 200×40 blocks. The FAST-C(2D) model used a coarse mesh with 100×20 blocks. Besides different discretization schemes have been used for spatial 1st order derivatives: Yusa (1983) chooses central weighting, while Holzbecher/Yusa (1995) prefer the upwind scheme.

There are only marginal differences in the FD-models concerning the steady state. Holzbecher/Yusa (1995) introduce a new dimensionless number Co:

$$ Co = \frac{g\Delta\rho}{\partial p / \partial x \big|_{top}} = \frac{kg\Delta\rho}{\mu v_{bound}} = \frac{kg\Delta\rho H}{\mu D} \frac{D}{v_{bound} H} = \frac{Ra}{Pe} \qquad (10.5) $$

which relates the free and forced convective regimes. Note that Pe denotes the dimensionless Péclet number as a combination of physical parameters with the characteristic length H - while the grid-Péclet number requires the grid spacing.

The following classification concerning the steady state of the Yusa geothermal model is valid for both the earlier and the later calculations.

Hydraulic gradient	Co=Ra/Pe	Steady state pattern
0.0	∞	one steady roll
1 %	23.06	two steady convection rolls
3 %	7.69	no steady state
5 %	4.69	no convection

Table 10.3: Characterization of free and forced convection

The classification of Table 10.3 was published first by Holzbecher/Yusa (1995) and is basically confirmed by the results of Yano (1989) and Springer (1995). Besides some details mentioned above there is one major exception for which calculated flow is different: in Springer's model a one-eddy regime is produced for the 1% hydraulic gradient. Note that the values in the second column slightly differ from those given in Holzbecher/Yusa (1995).

Yano (1989) and Holzbecher/Yusa (1995) discuss the effect of variable viscosity. Note that the viscosity change is approximately one order of magnitude for the considered range of temperatures (see chapter 2, equations (2.5) and (2.6)). Both models show a qualitative change for the steady state in the low gradient case. While an ascending flow regime is established along the right boundary, the flow remains in downward direction for strong viscosity changes. The pattern of motions basically remains as it is in the beginning (Fig. 10.3).

In the $I=0.03$ case the downflow pattern on the right side of the model is also not changed, when variable viscosity is taken into account. The main features of flow like those shown in Fig. 10.7. Note that for constant viscosity the computed flow patterns were different for cases $I=0.01$ and $I=0.03$.

For the strong gradient case with $I=0.05$ no substantial effect due to variable viscosity can be observed.

Finally, it should be noted that all results calculated by Springer (1995) have been obtained under consideration of variable viscosity.

There are differences concerning the transient development of the flow patterns in the different simulations, but they are not discussed here. Some remarks can be found in Holzbecher/Yusa (1995).

10.3 Sensitivity Analysis

Some new calculations are presented using FAST-C(2D) code for the free convection situation $(I=0)^2$. Holzbecher/Yusa (1995), as mentioned above, use the upwind scheme (see chapter 4.1). This scheme has the advantage delivering an unconditionally stable algorithm. On the other hand, there is the disadvantage that numerical dispersion is produced as well when grid-Péclet numbers are high. If the steady situation of the free convection case is taken, grid-Péclet numbers are 3.45 in x-direction and 5.29 in z-direction.

There is numerical dispersion in the results published by Holzbecher/Yusa (1995) - introduced by the truncation error as described in chapter 4. On the other hand, there is a qualitative correspondence with the output that Yusa (1983) obtained with a finer grid. Runs with FAST-C(2D) for the coarse grid choosing central weighting became unstable. This could be expected from the fact that grid-Péclet numbers are still above 1.

FAST-C(2D) is applied in a new simulation series to study the effect of 1st order discretization techniques, spatial grid and timestep resolution. The grid has to be selected at least as fine as the one used in the original publication of Yusa (1983) in order to do that. Central (CIS) and backward in space (BIS) discretization schemes are chosen to treat 1st order terms in the transport equation (see chapter 4, equations (4.4) and (4.5)). Timesteps are varied as well. Timestep size changes

---

[2] The input data for the model can be found on CD-ROM in the files *Yusa.WTX*, *Yusa1.WTX, Yusa2.WTX, Yusa3.WTX.*

using alternatively 12 or 18 steps for the first time period of 200 years. The
number of steps in the further timeperiods are listed in the following table.

Time=0	to 200 a	to 1000 a	to 2000 a	to 3000 a	to 4000 a	to-5000 a
Number of	12	24	24	24	24	24
timesteps	18	36	36	36	36	36

Table 10.4: Timesteps for model runs to simulate the free convection case

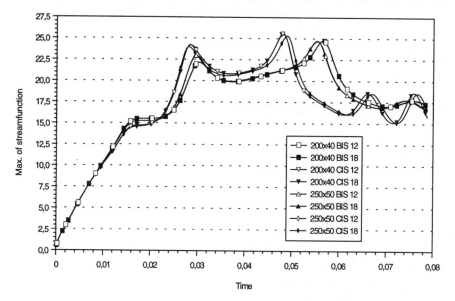

Fig. 10.9: Transient development of streamfunction using FAST-C(2D)
200×40 and 250×50 indicate number of blocks in vertical and horizontal direction;
BIS and CIS indicate 1st order term discretization: backward and central in space;
12 and 18 indicate timestep size: 12 and 18 steps in first timeperiod; see Table 10.4

Fig. 10.9 shows the transient development of the maximum streamfunction for
the situation without hydraulic gradient. The results do not change, if smaller
timesteps are chosen. Thus timesteps are fine enough. Remarkably it is obviously
not necessary to fulfill the Courant criterion: Courant numbers are above the limit
(Table 10.5).
Major differences are due to the 1st order discretization schemes. Results using
CIS are different from those using BIS. Local minima or maxima are reached at
different times (Fig. 10.9). CIS-outputs show fluctuating regimes at the end of the
simulated timeperiod. In contrast BIS-simulations run into a steady state. This is
possibly caused by numerical diffusion. Correction schemes for BIS-discretization
could not be applied because grid-Péclet-numbers are still too high (Table 10.5).

There is still a dependence on the spatial resolution. Results for the 250×50 grid still differ from those for the 200×40 grid. The temporal development is almost the same for runs with different timesteps. But an increasing shift of the time axis can be observed. The fine grid solution runs behind, when the CIS method is applied. The coarse grid solution runs behind for the upwind scheme.

Time	CIS (central in space)				BIS (upwind scheme)			
	Pe_x	Pe_z	Cou_x	Cou_z	Pe_x	Pe_z	Cou_x	Cou_z
.0032	1.75	0.72	1.53	0.63	1.68	0.73	1.47	0.64
.0094	1.93	1.53	3.38	2.67	1.91	1.52	3.36	2.67
.0158	1.91	1.47	3.34	2.57	1.89	1.59	3.31	2.78
.0236	1.93	1.68	4.23	3.67	1.87	1.18	4.08	2.58
.0315	1.84	1.72	4.02	3.75	1.82	1.87	3.98	4.09
.0394	1.79	1.50	3.92	3.28	1.77	1.74	3.86	3.81
.0472	1.77	2.08	3.86	4.55	1.73	1.46	3.79	3.19
.0551	1.70	2.08	3.73	4.55	1.70	2.10	3.72	4.81
.0630	1.70	2.02	3.72	4.42	1.65	2.09	3.61	4.57
.0708	1.69	1.89	3.69	4.14	1.65	2.14	3.60	4.68
.0787	1.69	2.19	3.69	4.78	1.64	2.09	3.59	4.56

Table 10.5: Maximum grid-Péclet and Courant numbers for two simulations of the free convection case; the grid consists of 250×50 blocks; 18timesteps are used per timeperiod

It can be concluded that the grid resolution of 250×50 is not yet sufficient. But the changes are relatively small. No further refinement is made here. Interested readers may take the example input file from CD-ROM and run further models using FAST-C(2D).

10.4 Other Geothermal Systems

The subject of geothermal flow was treated in this chapter in a quite restricted manner. The problem of Yusa examining the interaction of potential and geothermal flow under natural conditions was given exclusive attention. Other systems are relevant and have been discussed in the literature.

Several publications discuss the situation in which the hot temperature region extends across the entire lower boundary (Sorey 1978, Cheng 1978, Garg/Kassoy

1981). This is in fact the same situation as in Bénard convection. Several points which are discussed in publications on geothermal systems, are treated in this book in chapters on Bénard convection. One of these subjects is the onset of convection which can be found here in chapter 5. Another example is the influence of variable of viscosity on fluid motions (chapter 6.2).

Various aspects lead to a distinction of various types of geothermal reservoirs. Geothermal systems can be divided into those which are in their natural state and those which are exploited. There are several different technologies to exploit a geothermal reservoir. Heat pumps extract heat only. When hot water is pumped to the surface, heat is usually gained using a heat exchanger. In some applications cooled water is recharged into the subsurface again.

Geothermal resources can be classified in low-temperature and high-temperature systems. With respect to temperature range a world-wide accepted classification does not exist. In regions of volcanic activity temperatures reach several hundreds degree Celsius, while in other regions 'hot' water of 100°C can hardly be found - Germany belongs to the latter class.

Geothermal reservoirs may be located in fractured rock and require a different modeling approach than for the porous medium case.

Another distinction is between single-phase and two-phase systems. While in the former only fluid water or gaseous steam need to be considered, there is water and steam in the latter situation. Two phase systems are generally much more complicated than single phase systems. In geothermal systems the phase transition from water to steam or vice versa complicates the analytical treatment additionally. In regions of geothermal activities the two phase reservoirs are quite common. Despite of it's importance the treatment of multi-phase flow is beyond the scope of the book.

1D-, 2D and 3D-models can be found in the literature on geothermal systems. With 1D and 2D models often a hypothetical situation is studied from which conclusions can be drawn for the behavior of geothermal systems in general. Examples for this type of simulation are presented by Donaldson (1970), Cheng/Lau(1975), Kassoy (1975), Ziagos/Blackwell (1986).

2D and 3D models are set up for several application case studies where site-specific concepts and parameters are introduced. Only some examples are listed here finally. The Wairakei system in New Zealand has been modeled by Mercer/Pinder (1973) and Blakeley/O'Sullivan (1982); the Mesa system in California by Morris/Campbell (1981); the Baca geothermal field in New Mexico by Bodvarsson e.a. (1982); the Cerro Prieto geothermal filed in Mexico by Lippmann/Bodvarsson (1983); the Lassen system in California by Ingebritsen/Sorey (1985); the Yufuin system in Japan by Holzbecher (1993); the Oguni geothermal field in Japan by Nakanishi e.a. (1997) and Pritchett/Garg (1997).

11 Saltwater Intrusion (Henry's Example)

11.1 Problem Description

The intrusion of saltwater from a saltwater reservoir into a connected freshwater aquifer is a well-known problem not only on the seashore. Fig 11.1 shows the situation on the German North-Sea coast. It is a severe problem where coastal regions are highly populated, especially in the vicinity of big cities. The situation is worsened by increased pumping of groundwater for irrigation, drinking water or recreational purposes.

Saltwater intrusion is a phenomenon in coastal regions where the salinity of the sea 0or some other surface water body is higher than the salt content of the natural groundwater. Even under natural conditions without any anthropogenic activity the water table fresh water from the aquifer flows into the sea, while in greater depth saline water penetrates into the pore space. The saltwater wedge at the bottom of an phreatic aquifer may move long distances against the natural gradient at the water-table. The driving force for this phenomenon is the higher density of saline fluid in comparison to fresh water.

The penetrating saltwater is stopped at some point by the fresh water flow, which mostly moves towards the shore following a natural gradient. A transition zone is built between the saline and fresh water parts. It is characterized by

- brackish water with intermediate salinity
- water with salinity slightly below sea water level that enters from the sea into the lower part aquifer
- water with slightly increased salinity that flows towards the sea

In some cases a very small mixing zone can be observed - idealized as *'sharp interface'*. The mixing is caused naturally by dispersion and tidal motions. This in most cases leads to the emergence of a transition zone in which salt concentrations increase gradually up to sea water salinity level.

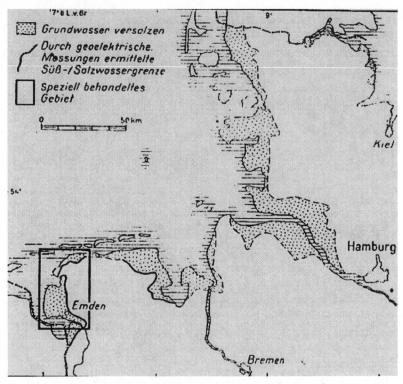

Fig. 11.1: Saltwater intrusion along the German North Sea shoreline; dotted areas indicate problem zones (Richter/Flathe 1954)

Increased pumping of groundwater reduces the natural gradient of fresh water. The direction of the flow is even reversed in the direct vicinity downstream of the wells. The force limiting saltwater intrusion is smaller and saline fluid proceeds further. If the pumping regime is nearly constant, a new equilibrium is approached.

The new steady state has increased salt concentrations in the subsurface. Those fresh water wells nearest to the shore may pump brackish or even saline water under the new conditions. It may become necessary to close some facilities when the salt content of the produced water is too high for certain purposes.

The situation will not meliorate if the total amount of pumped water is not reduced, or if other actions are initiated to prevent further intrusion. Aside from reduced pumping, several actions can be taken to reach an acceptable situation. The installation of wells for artificial recharge has become the preferred technique and is used in many parts of the world. Some US-studies report on *'intrusion barriers'* or *'fresh water barriers'* of surface water (Welsh 1989, Richardson 1989). The technology is applied in large scale in the dunes in the Netherlands of North Sea coast (Piet/Zoeteman 1985, Peters 1989). Regional numerical models may help find a suitable groundwater management strategy with *'redistributed abstraction'* to cope with the problem (Spink/Wilson 1987). The construction of a subsurface wall to prevent the saltwater wedge from further migration is studied by Jovkov (1987).

11.2 Sharp Interface Approach

Saltwater intrusion was described already in the 19th century. Badon-Ghyben (1889) cites several Dutch sources, for example: '*Het valt niet te betwijfelen, of het zeewater doordringt onze bodem op eene zekere diepte, die met den afstand toeneemt*'[1]. Independent from each other, Badon-Ghyben (1889) and Herzberg (1901) found that the depth of the saltwater is related to the density difference between the fluids[2]. They derived a formula which relates the fresh water head above sea level h_f and the distance of the salt-freshwater interface below the sea level z_s (see Fig. 11.2):

$$z_s = \frac{\rho_f}{\Delta\rho} h_f \qquad (11.1)$$

Sea water and freshwater have the ratio of $\rho_f/\Delta\rho \cong 1000/25 = 40$. Using different values for sea water density Badon-Ghyben calculated the ratio as 42 and Herzberg as 37. The derivation of the formula is based on the assumption of an hydrostatic equilibrium below and above the interface. The weight of two vertical water columns is compared: one with a length of z_s containing salt water only and the other with a length of z_s+h_f containing freshwater (see Fig. 11.2). The weight of both columns is equal at the interface:

$$\rho_s z_s = \rho_f (z_s + h_f) \qquad (11.2)$$

Solving (11.2) for z_s leads to equation (11.1). Hubbert (1940) writes: '*It is a relationship of great usefulness but also whose use entails considerable danger of error, because it is not correct*'.

The validity of the rule has been confirmed using nowadays exploration technologies. Wolff J. (1997) reports a maximum thickness of 80 m for the fresh water lens and an elevation of up to 2 m above mean sea level for groundwater of the German North Sea island Nordeney. The Ghyben-Herzberg formula can only be applied in phreatic aquifers with measurable rise of the watertable.

Approaching the sea, according to (11.1) the interface should rise up to the shoreline. But this cannot be observed in nature. A groundwater outflow face can

---

[1] There is no doubt that seawater fills the soil at a depth that increases with distance [from the sea - E.H.]

[2] The historical review by Reilly/Goodman (1985) reveals that the stated relationship has been used already by scientists before Badon-Ghyben and Herzberg.

mostly be found instead. Bear/Veruijt (1987) note that the assumption of hydrostatic conditions in the freshwater zone can be dropped and that hydraulic heads on both sides of the interface may be related instead of hydrostatic heads.

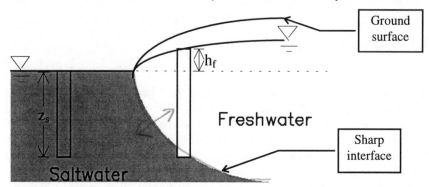

Fig. 11.2: Two hypothetical columns to explain saltwater intrusion

Another assumption of the Ghyben-Herzberg formula is the existence of a sharp interface separating both regions. In reality there is no mechanism preventing salt and fresh water from mixing. The transition zone may be quite small: then the application of a sharp interface model may be justified. But the mixing zone with brackish water often extends some distance and the sharp interface approach is of no use.

There are other limiting factors for the application of the Ghyben-Herzberg approximation. The author has been involved in a study on sea water intrusion into the Nile delta aquifer. There is a mean gradient of 11 cm/km over a distance of approximately 100 km from the Mediterranean. Using Ghyben-Herzberg as a rule of thumb, groundwater with sea water salinity should be expected even in the middle of the delta, about 100 km south of the Mediterranean. Measurements and model results show that this is not true (Holzbecher/Baumann 1994, Baumann 1995, Arlt 1995).

Beyond other restrictions, the applicability of equation (11.1) is obviously a question of scale. It does not work in the direct vicinity of the shore (<10m), but it is possibly applicable in the 100 m/ 1 km range for which it was designed in the original publications of Badon Ghyben (1889) and Herzberg (1901). Kohout/Klein (1967) showed for the Biscayne aquifer that the Ghyben-Herzberg formula overestimates saltwater penetration. It seems that the validity must be doubted generally for long distances.

Essaid (1986) stresses the point that the Ghyben-Herzberg rule is based on the assumption, *'that the salt water flow zone can instantaneously adjust to changes in the fresh water zone'*. The transient reaction of the saline fluid is neglected, but it is especially relevant when water is pumped. Essaid concludes: *'The duration of the period over which the system does not behave as predicted by the Ghyben-Herzberg principle can be as much as 10 to 20 yrs'.*

11.3 Henry's Example

It is the hypothesis of Cooper (1959) that the conditions in coastal aquifers are not static: '*salt water... flows perpetually in a cycle from the floor of the sea into the zone of diffusion and back to the sea*'. One relevant mechanism is the mixing of the fresh and saline waters. The circulating motion can be explained taking variable density into account. In the words of Cooper (1959): '*the diluted sea water, having become less dense than the native sea water, rises along a seawater path. The resulting circulation is analogous to the circulation in thermal convection, differing from it only in that changes in density are produced by changes in concentration rather than by changes in temperature*'.

It was Henry (1960) who proposed an analytical description of the situation. He tackled the mixing of fluids with different densities and is led to a quantitative description of the dynamic balance between fresh and saline fluids in coastal aquifers. The flow pattern of saltwater intrusion is steady-state.

The considered (model-) area is a 2D-vertical cross-section through the aquifer, where one vertical boundary is situated near the sea shore. Henry starts with Darcy's law, the continuity equation, and a steady state transport equation (compare chapter 3). In his analysis he introduces the streamfunction ψ and normalized concentrations θ. He makes simplifications similar to those presented in chapter 3 and comes up with the following system of partial differential equations:

$$a \cdot \nabla^2 \psi = \frac{\partial \theta}{\partial x} \qquad b \cdot \nabla^2 \theta - \frac{\partial \psi}{\partial z}\frac{\partial \theta}{\partial x} + \frac{\partial \psi}{\partial x}\frac{\partial \theta}{\partial z} = 0$$

$$\text{with} \qquad a = \frac{\mu Q}{kg\Delta\rho H} \qquad \text{and} \qquad b = \frac{D}{Q} \qquad (11.3)$$

where Q denotes the net freshwater discharge per unit length of the seashore. The transformation of the streamfunction $\psi \rightarrow \Psi = \psi D/Q$ results in the system of differential equations which is implemented in the FAST-C(2D) code:

$$\nabla^2 \Psi = \frac{kg\Delta\rho H}{\mu D}\frac{\partial \theta}{\partial x} \qquad \nabla^2 \theta - \frac{\partial \Psi}{\partial z}\frac{\partial \theta}{\partial x} + \frac{\partial \Psi}{\partial x}\frac{\partial \theta}{\partial z} = 0 \qquad (11.4)$$

The coefficient in the first equation is the Rayleigh number, as known from simulations of convective motions. Henry (1964) provides solutions for his model when the dimensionless constants a and b take the values shown in the first two columns of Table 11.1. The corresponding Rayleigh number is given in the third column.

$a=Q\mu/kg\Delta\rho H$	$b=D/Q$	$Ra=1/(a\cdot b)$
0.263	0.1	38.0
0.450	0.1	22.2

Table 11.1: Input values for the Henry saltwater intrusion test problems

Henry introduces a concept for diffusivity which contains more than molecular diffusion. Mixing effects due to flow inhomogeneity are included as well; D is an *effective diffusivity*. Dispersivity is usually proportional to velocity and therefore changes in space. Huyacorn e.a. (1987) propose velocity-dependence with dispersion lengths as given in Table 11.4.

At the seaside the tides may also influence the dispersion due to temporal changes in the flow regime. Dei (1978) found in an experimental set-up that amplitude and frequency of oscillations have an influence on the effective diffusivity.

Jennings (1984) studied the influence of tidal oscillations in a numerical study. Input values in that study were specified to be consistent with Henry (1960) - see Table 11.3. In addition a fluctuating boundary condition for hydraulic head was introduced on the seaside boundary. It was observed on the numerical results that with every tidal cycle the dispersed front proceeds further into the aquifer. With the 4th tidal cycle the 0.1 isochlor reaches the left boundary of the model region. The results of the calculations indicate that the influence of tidal motions may be dramatic. Numerical models could be used to quantify the effect of tidal oscillations on the extension of the dispersion front. The publication of Jennings (1984) shows clearly that in the Henry problem for such a study the horizontal extension is too small.

The boundary conditions as specified by Henry are given is the first three columns of Table 11.2.

Boundary	θ	ψ	Ψ
left	Dirichlet: $\theta=0$	Neumann: $\partial\psi/\partial x=0$	Neumann: $\partial\Psi/\partial x=0$
right	Dirichlet: $\theta=1$	Neumann: $\partial\psi/\partial x=0$	Neumann: $\partial\Psi/\partial x=0$
top	No flow: $\partial\theta/\partial z=0$	Dirichlet: $\psi=1$	Dirichlet: $\Psi=10$
bottom	No flow: $\partial\theta/\partial z=0$	Dirichlet: $\psi=0$	Dirichlet: $\Psi=0$

Table 11.2: Boundary conditions of the Henry saltwater intrusion test problems

There have been some critical remarks in several publications on the constellation as specified by Henry. The two main points which I want to mention are the unrealistic boundary condition on the seaside, and the approach of a constant diffusivity.

The boundary condition which Henry requires for the seaside has been questioned by many researchers. Huyacorn e.a. (1987) propose a different boundary condition, where the upper part of the boundary is open to outflow and the salt water boundary is used for the lower part of the aquifer. For dimensionless variables θ and z' measured upward from the bottom this reads:

$$\partial \theta / \partial x = 0 \qquad \text{for } x=L \text{ and } z'>0.8$$
$$\theta = 1 \qquad \text{for } x=L \text{ and } z'\le 0.8 \qquad (11.5)$$

This boundary condition has been adopted by many researchers including Frind (1982) and Putti/Paniconi (1985). It is surely more realistic, because the original formulation foresees no outlet for freshwater. Because the upper model edge is a streamline, there is no outflow on the top. And because of the $\theta=1$ condition at the right hand boundary, the concentration of the mixed fluid necessarily has to increase up to seawater level before the outflow from the system. On the other hand, when the boundary condition (11.5) is used, the heights of the inflow and outflow parts beneath the seashore are prescribed.

11.4 Modeling Saltwater Intrusion

Various model types have been used to tackle the problem of saltwater intrusion. Models can be divided into those based on the sharp interface assumption and those which use the concept of miscible displacement.

Early approaches for salt water intrusion used the sharp interface approach (see chapter 11.2). They are based on the assumption that saline and fresh water do not mix and that a sharp interface between both phases is formed. There is no flow across the interface. The Ghyben-Herzberg formula (11.1) stems from the very same idea (see above).

The general formulation of the mathematical problem becomes easier with the assumption of a sharp interface. In 2D the shape of the interface curve (which is 1D) is required. This reduces the problem dimension and simplifies problem and solution techniques (Wilson/Sa da Costa 1979, 1982, Bear/Veruijt 1987). Some researchers restrict the problem to tracing the toe-point of the saltwater front.

Essaid (1986) makes a distinction between single-fluid and two-fluid sharp interface approaches. In the single-fluid case a differential equation for fresh water hydraulic head h_f is solved, where h_f is given by:

$$h_f = \frac{p}{\rho_f g} - z \qquad (11.6)$$

The depth of the interface can be computed easily using the Ghyben-Herzberg formula (11.1) in a post-processing module. *'The simplifying assumption is that the salt flow zone can instantaneously adjust to changes in the fresh water zone'* states Essaid (1986). An improvement is given when the salt water head

$$h_s = \frac{p}{\rho_s g} - z \qquad (11.7)$$

as introduced by Hubbert (1940) is considered as well. The interface position z can then be determined from the condition that h_s and h_f are in equilibrium:

$$z = \frac{\rho_f}{\Delta \rho} h_f - \frac{\rho_s}{\Delta \rho} h_s \qquad (11.8)$$

The treatment of saltwater intrusion with a sharp interface model is surely an improvement in relation to Ghyben-Herzberg's rule of a thumb (see chapter 11.1). Sharp interface models were certainly more appropriate when computers capabilities were quite restricted. Speed and storage capacity have increased so much that coupled flow and transport models can be run today on personal computers.

The assumption of non-mixing fluids is physically unjustified and all cases build a differently wide transition zone. The sharp interface is replaced by a mixing zone of fresh and saline water. The main reason for the introduction of the sharp interface assumption is that it is mathematically easier to handle. Henry (1960), with his above mentioned conceptual model, was the first one, who preserved the miscible displacement approach despite its considerable difficulty.

Several models have been presented in scientific literature using the coupled flow and transport equations. Different numerical approaches are used. A steady state is often calculated, while the transient development within the system requests more computational efforts. The models can be classified in five dimension sets: 1D, 2D-vertical, 2D-horizontal, 2D-horizontal for leaky aquifers, 3D. Various formulations are used: the flow variable can be represented by total pressure, hydraulic head or streamfunction. Various discretization approaches have been used: finite elements, finite differences, finite volumes, boundary elements. Moreover there are various different strategies to solve linear and nonlinear systems (Pinder/Sorek 1998).

Most codes are written for the saturated case; few allow the choice of unsaturated conditions. Codes differ in the treatment of sorption: a single component representing salinity is usually used; few include chemical reactions. Moreover isothermal conditions are assumed by the majority of researchers.

Hill (1988) provides a comparison between sharp-interface and convection-dispersion models. For different types of aquifers, he finds that *'the sharp-interface model may produce an estimate of the location of saltwater in coastal aquifers that is further landward than the convective-dispersive model'*. Compared with real data, the interface is much too far inland.

A good overview on the state of the art in the mid-80s is given by Reilly/Goodman (1985). They write in conclusion: '*We have, however, reached a point in which most of the important factors are known and the field problem can be addressed quantitatively*'. More modelers attempted to include a better representation of the interacting processes in the last decade. Recent publications show that the mixing concept dominates and the sharp-interface approach is less often applied.

The following modeling of saltwater intrusion does not address the general 3D problem. The studies are restricted to 2D vertical cross-sections. Henry's example and a closely related case are treated in detail.

11.4.1 Henry's Example

Henry (1960) provides what he calls an '*analytical solution*' for the problem. In fact it is a numerical solution using the spectral method (Peyret/Taylor 1985, Canuto e.a. 1987). He starts with double Fourier series of the following type:

$$\Psi = \sum_{m=1}^{\infty} \sum_{n=1}^{\infty} A_{m,n} \sin(m\pi z) \cos(n\pi x / L)$$

$$\theta = \sum_{m=1}^{\infty} \sum_{n=1}^{\infty} B_{m,n} \sin(m\pi z) \cos(n\pi x / L)$$

(11.9)

The coefficients $A_{m,n}$, $B_{m,n}$ are unknown. The use of the Galerkin technique, already mentioned in chapters 5 and 7, enables the derivation of a set of nonlinear equations for the unknowns Ψ and θ. Henry uses a numerical procedure to calculate a finite number of these coefficients: 38 of $A_{m,n}$, and 40 of $B_{m,n}$. A detailed reconstruction and evaluation of the method, as presented by Ségol (1994), shows that the calculations are basically correct. Although some oscillations in the isohalines can be observed locally, the general outline of the solution does not change, even if 100 terms of the $B_{m,n}$ are considered.

Henry's constellation of saltwater intrusion has become the favorite test-case for all codes designed to model this phenomenon. Pinder/Cooper (1970), Lee/Cheng (1974), Ségol e.a. (1975) treated the problem using FE-codes; Frind (1982) slightly varies the boundary conditions of the original definition; Jennings (1984) studies the influence of tidal oscillations on dispersion (see Table 11.3); Sanford/Konikow (1985) demonstrate that Henry's case can be modeled by an extended version of the code of Konikow/Bredehoeft (1978) using the method of characteristics (MOC) for the simulation of advective transport; Huyacorn e.a. (1987) check their 3D-multiple aquifer model using physical units for their parameters as they are reproduced here in Table 11.4. They also include velocity dependent dispersion. As noted above Henry's original formulation made use of a effective diffusivity that is constant throughout the entire domain.

Parameter	Symbol	Value
Length	L	2 m
Height	H	1 m
Permeability	k	$1.336 \cdot 10^{-9}$ m^2
Porosity	φ	0.35
Dynamic viscosity	μ	$1.334 \cdot 10^{-4}$ kg/m/s
Fresh water salinity	c_0	0 kg/m^3
Sea water salinity	c_1	24.5 kg/m^3
Fresh water inflow	Q	$6.6 \cdot 10^{-5}$ m^3/s

Table 11.3: Input data for Henry test-case with physical units according to Jennings (1984)

Parameter	Symbol	Value
Length	L	200 m
Height	H	100 m
Density difference	$\Delta\rho$	25 kg/m^3
Hydraulic conductivity	K	1 m/d
Porosity	φ	0.35
(Effective) Diffusivity	D	$6.6 \cdot 10^{-2}$ m^2/d
Longitudinal dispersivity	α_L	3.5 m
Transversal dispersivity	α_T	3.5 m
Fresh water inflow velocity	v	$6.6 \cdot 10^{-3}$ m/d

Table 11.4: Input data for Henry test-case with physical units according to Huyacorn e.a. (1987)

Voss/Souza (1987) remark that the original problem formulation and the formulation given in physical units are not equivalent. Henry in his early publication obviously used a diffusivity for the porous medium D^*, which is connected with the diffusivity in a fluid phase D by:

$$D = D^* / \varphi \qquad (= 18.86 \cdot 10^{-6} \, m^2 /s) \qquad (11.10)$$

In some publications D^* is referred to as *effective diffusivity* (Cussler 1984, Fan e.a. 1997).

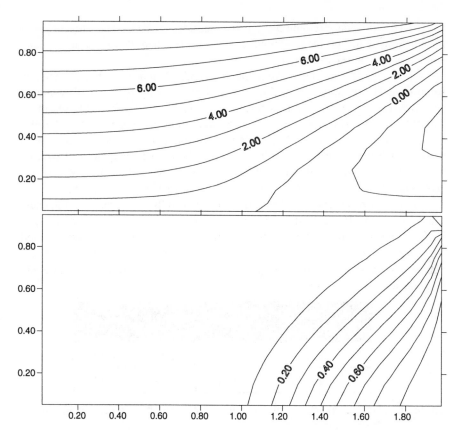

Fig. 11.3: Steady state modeling for the Henry example; the upper graph depicts
streamlines, the lower graph normalized salinity; numerical parameter
details can be found in the text

Despite criticism, Henry's specification has been used as a standard test-case
for codes treating variable density flows since begin of the 80s. It has been used by
INTERA (1979) for SWIFT; by Voss/Souza (1987) for SUTRA (Voss 1984); by
Holzbecher (1991a) for an early version of FAST-C(2D); by Kolditz (1994) for
FEFLOW; by Strobl/Yeh (1994) for 2DFEMFAT; by Oldenbourg/Pruess (1995)
for TOUGH2; and by Bastian e.a. (1997) for UG. A Eulerian-Lagrangian approach
has been suggested by Galeati e.a. (1992). Putti/Paniconi (1995) take the Henry
test-case to compare different linearization approaches. The test-case is chosen by
Senger/Fogg (1990). as a test example for the application of the streamfunction
formulation.

Fig. 11.3 shows the results of a new modeling run with FAST-C(2D)[3]. The grid
spacing is equidistant with 40 blocks in horizontal and 10 blocks in vertical
direction. The central scheme (CIS) is selected for the discretization of 1st order

---

[3] The input data for the model can be found on CD-ROM in the file *Henry.WTX*

terms. The boundary condition on the seaside is $\theta=1$. The dimensionless formulation of the problem is used with $L=2$, $H=1$ and $Ra=38$. The boundary conditions are chosen as listed in Table 11.1.

Comparison of the output from the FAST-C(2D)-model with the analytical solution of Henry shows some differences. The toe-point of the 0.5-isohaline is usually used as a reference. This point can be found farther from the seaside in Henry's solution. This deviation has been stated by many modelers (Ségol 1994). She presents a recalculation of Henry's formulation and finds the toe-point of the 0.5-isohaline very near to 1.4 on the abscissa. This coincides quite well with the results presented above. A very similar output has been obtained by Voss/Souza (1987) using SUTRA code with diffusivity $D=18.86 \cdot 10^{-6} \text{ m}^2/\text{s}$ (compare (11.10)).

A short test series shows the sensitivity of the results on numerical parameters. The original 40×10 equidistant grid was refined twice in order to reduce grid-Péclet numbers. The first refinement increases the number of blocks in both coordinate directions to a 80×20 grid. The second refinement doubled the grid numbers in horizontal direction to a 160×20 grid. The resulting grid-Péclet numbers are given in Table 11.4.

Grid-Péclet-numbers	40×10	80×20	160×20
Pe_x	2.56	1.58	0.82
Pe_z	1.18	0.68	0.74

Table 11.5: Henry's example - maximal grid Péclet-numbers in dependence on grid spacings

The grid Péclet numbers have been obtained using the upwind scheme for the discretization of the 1st order terms in the differential equation. Runs of the model have also been performed using the central scheme. Fig. 11.4 shows a comparison on the contours of normalized concentration for the values from 0.1 to 0.9 (step size 0.1) for all of the noted variations.

The differences between the different model runs are marginal. It is remarkable that the central scheme run for the coarse grid converges at all. It yields good results although the grid Péclet criterion is not fulfilled. Differences between upwind scheme and central scheme become smaller and nearly vanish for the finest grid. Runs with correction of the truncation error have been made for both coarse grids and both gave totally different results. Apparently this option can only be used if the grid Péclet-criteria are fulfilled.

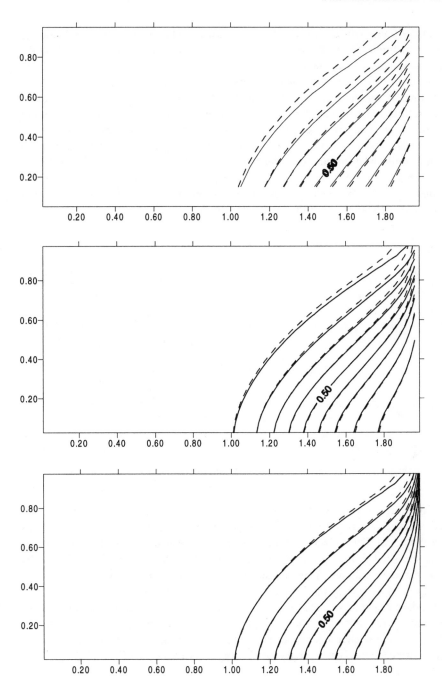

Fig. 11.4: Steady state simulations for Henry's example with FAST-C(2D); dependence on grid: top: 40×10, center: 80×20, bottom: 160×20 blocks dependence on discretization: broken lines: upwind, else: central scheme

In the finest grid differences between upwind scheme and central scheme are hardly distinguishable. It follows that the numerical dispersion introduced by the upwind scheme is very small.

Test have been made to find out if the obtained solution for the Henry problem is unique. A transient simulation for the problem has been initiated for that reason. The results of two runs have been compared. One run starts with an aquifer entirely filled with fresh water. The second run starts with an aquifer filled with saline water. Both runs reached the same steady state. In the first run a dispersion zone is built up from the salt-water boundary (in the figures on the right side). In the second run fresh water gradually removes the salt-water from the inland boundary (left side). The steady state reached by both transient simulations coincides with the result of the steady state model.

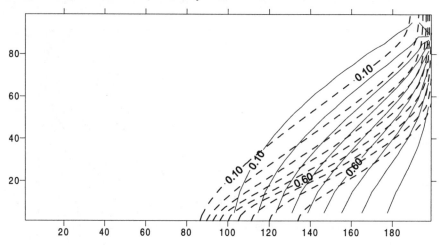

Fig. 11.5: Concentration contours for constant and velocity dependent
dispersivity

The effect of velocity dependent dispersion has also been checked. Longitudinal and transversal dispersivity have been chosen according to Table 11.3, as $\alpha_L = \alpha_T = 3.5$ m. For molecular diffusivity D a smaller and more realistic values has been specified. It turned out that there are problems for low values of diffusivity D. A parameter stepping method reached low values as $D = 1 \cdot 10^{-9}$ m^2/s.

Fig. 11.5 compares the isohalines of a constant $D = 5 \cdot 10^{-9}$ m^2/s and a velocity dependent dispersion tensor. The velocity dependent case has the toe point of the contours further inland, while the slope of the curves becomes much steeper in the vicinity of the seaside boundary. The same observation has already been made by Huyacorn e.a. (1987) and Frind (1982) using other codes than FAST-C(2D).

11.4.2 Parameter Variation

The model of saltwater intrusion presented above goes back to Henry (1960).Two dimensionless parameters determine the system. One is the Rayleigh number Ra, which appears in the differential equation (11.4). The other parameter is the difference of the streamfunction $\Delta\Psi$ for the two horizontal levels.

Length L is another parameter appearing in the conceptional and numerical models. The intention here, however, is to study solutions which are independent of L. It is always possible to obtain an intrusion solution independent from L, if the system is chosen long enough.

Other dimensionless parameter combinations besides Ra can be used to describe the behavior of the system. A (physical) Péclet number has been introduced by Volker/Rushton (1982) as:

$$Pe = \frac{KH}{\varphi D}$$
(11.11)

Péclet numbers relate advective and diffusive processes and this is done by the given formula. Equation (11.11) is obtained by taking H as a characteristic length and K/φ as a characteristic flow velocity. The definition of the Péclet number is not unique. The following uses a form more convenient to the problem of saltwater intrusion.

Volker/Rushton (1982) tackle saltwater intrusion using dimensionless parameters, but one crucial parameter is not considered in the description. Fresh water inflow has a decisive importance for the penetration of the saltwater wedge. Several parameters could be used to quantify this crucial characteristic. Henry (1960) used a flow rate Q which is identical to $\Delta\Psi$ in the streamfunction formulation. The fresh water inflow velocity v_{in} could be taken as characteristic parameter.

The definition for Pe in equation (11.11) cannot describe the distance of intrusion. A convenient way to include fresh water inflow, is to use v_{in} as characteristic velocity:

$$Pe = \frac{v_{in}H}{D}$$
(11.12)

v_{in} can be expressed in terms of $\Delta\Psi$ and Q. The value of $\Delta\Psi$ is equal to the fresh water inflow Q. Thus Q is connected to the Péclet number according to:

$$Pe = v_{in}\frac{H}{D} = \frac{\Delta\Psi}{H}\frac{H}{D} = \frac{\Delta\Psi}{D} = \frac{Q}{D}$$
(11.13)

In the dimensionless system, where the diffusivity becomes 1.0 (see eq. (11.4)) thus holds:

$$Pe = \Delta\Psi$$
(11.14)

Another dimensionless parameter has already been defined by equation (10.5). The characteristic velocity for the saltwater intrusion case is the inflow velocity and thus the dimensionless number is given here by the definition:

$$Co = \frac{kg\Delta\rho}{\mu v_{in}} = \frac{kg\Delta\rho H}{\mu D}\frac{D}{v_{in}H} = \frac{Ra}{Pe} \qquad (11.15)$$

A parameter variation has been made using the FAST-C(2D) code in order to relate the distance which the salt water wedge penetrates into the aquifer to the input parameters. There are two reasonable definitions for the *penetration length*. One definition is given by the toe point of the 50%-isohaline in the steady state solution, more precisely: the distance between the seaside boundary and the point on the aquifer bottom where the salt concentration decreases by one half. In the following this distance is named P_{L1}.

A second definition can be made using streamfunction output. A limiting streamline separates two characteristic flow regions. In the upper region, streamlines pass through the whole system from inflow to the outflow boundary. In the lower region streamlines enter the system at the seaside near the bottom of the aquifer, turn and move back to the same boundary. The Ψ-level of the separating streamline is equal to the value specified for the bottom by the boundary condition. Fig. 11.3 portrays the separating streamline with the level 1.0. Penetration can thus be defined as the distance between the toe point of the separating streamline and the seaside boundary. This will be called P_{L2} in the following.

The two definitions are not equivalent. The numerical results show that in typical systems holds: $P_{L1} < P_{L2}$. In degenerated systems, where salt-water intrusion is mainly an effect of diffusion and advective inflow from the sea is marginal, the opposite relation holds: $P_{L1} < P_{L2}$. Note that both definitions are made for convenience in calculation from model output.

Both defined penetration lengths may be too small in practical applications. In field studies a definition of penetration length depends on the acceptable level of salinity increase. The position of the 10%-isohaline may often be the better choice in practice. Here the definitions are made to study the behavior of mathematical solutions and how output changes with a change of input parameters.

The purpose of the parameter variation is to find the basic dependencies between model input and output, i.e. to find out how intrusion changes when certain system input changes. The following two figures show the results of the parameter variation. The penetration lengths P_{L1} and P_{L2} are given as contours in the plane of dimensionless parameters. The dimensionless length unit is H, i.e. the contour to level 1.0 marks all cases where penetration length is equal to aquifer depth. The parameter variation study uses a set of selected combinations for 8 values of Ra and 4 values $\Delta\Psi$. The model did not converge for some combinations. Some of them had the salt-water wedge passing through the entire system. The results of these combinations have been omitted because the penetration length given above is not a reasonable definition under those circumstances.

The model length has been chosen as $L=4$. Penetration lengths below 3 can be taken as valid for the concerned purpose because the influence from the inflow boundary otherwise cannot be neglected.

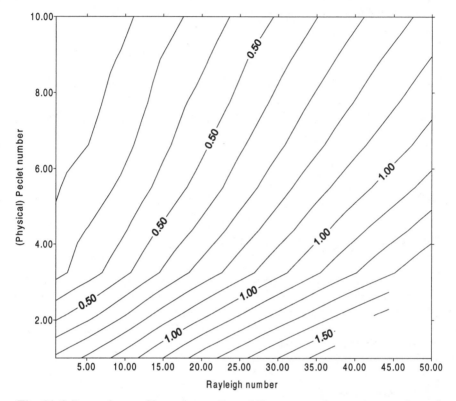

Fig. 11.6: Dependence of intrusion on Ra and Pe; portrayed are contours of equal penetration length P_{L1} (toe-point of 50%-isohaline) given in units of H

Fig. 11.6 depicts the dependency of penetration lengths P_{L1} on the dimensionless numbers Ra and Pe. Contours represent parameter combinations resulting in the same strength of intrusion if measured by P_{L1}. Pe represents the strength of the fresh water entering the system from one side. Penetration length decreases with increasing Pe. On the other hand Ra represents the strength of the density effect; increasing Ra strengthens intrusion. The qualitative behavior is as expected - nevertheless, the figure provides a quantitative estimation which may serve for cases lacking a detailed model.

A more instructive figure results when penetration length is presented in the Ra-Co-plane. Contours of P_{L2} from the parameter variation study are given in Fig. 11.7 and show a pattern of almost horizontal graphs. The result allows the formulation that penetration length is mainly a function of the dimensionless number Co.

This is not as amazing as might be thought at first. As noted above, *Ra* describes the strength of the (density-driven) buoyancy force driving penetration of saline water from the seaside. The force of the inflowing water acts from the other side and is represented by *Pe*. The dimensionless number *Co* as the ratio of both forces should therefore provide a good estimate for the strength of intrusion.

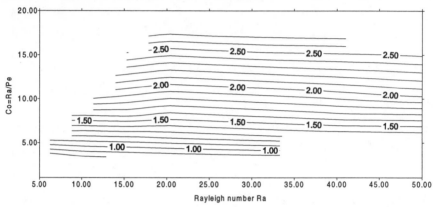

Fig. 11.7: Dependency of intrusion on *Ra* and *Co*; portrayed are contours of equal penetration length P_{L2} (toe point of streamline for level $\Psi=0$) given in units of *H*

Beyond this qualitative interpretation, Fig. 11.7 gives a quantitative estimation for the penetration length if some basic parameters are known. From the figure the penetration can be taken only for a certain parameter range. For parameters in a special application case outside that range, FAST-C(2D) users can easily set up an own study using their own data.

11.4.3 Layered Aquifers

Inouchi/Kakinuma (1992) presented a study on salt-water intrusion in a layered aquifer system. A 2D-model is set up to study the sensitivity of results due to the natural groundwater gradient, hydraulic conductivities of the layers, layer thickness and diffusivity. A schematic view of the conditions is given in Fig. 11.8.

Inouchi/Kakinuma use dimensionless parameters for the sensitivity analysis. They derive these parameters taking *d* as unit length and K_1 as unit velocity. Using the same transformations as in chapter 3 the steady state of the system can be described using the following differential equations:

$$\frac{\partial}{\partial x}\left(\frac{1}{k_z}\frac{\partial\Psi}{\partial x}\right)+\frac{\partial}{\partial z}\left(\frac{1}{k_x}\frac{\partial\Psi}{\partial z}\right)=-\frac{K_1\Delta\rho}{\mu}\frac{\partial\theta}{\partial x}$$

$$\frac{\partial}{\partial x}\left(D_x\frac{\partial\theta}{\partial x}\right)+\frac{\partial}{\partial z}\left(D_z\frac{\partial\theta}{\partial z}\right)-v_x\frac{\partial\theta}{\partial x}-v_z\frac{\partial\theta}{\partial z}=0$$

(11.16)

The following demonstrates some effects are using FAST-C(2D) as modeling code. The input data are listed in Table 11.6.

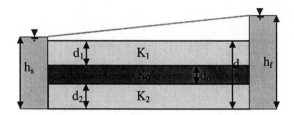

Fig. 11.8: Conceptual model for a study of intrusion into a layered aquifer system

Longitudinal and transversal dispersivity are set to zero. Diffusivity is assumed to be homogeneous. The permeability of the semi-permeable intermediate in the model layer is changed: K_0/K_1=1/5, 1/50 and 1/500. The conductivity of the bottom layer was chosen to be identical to the conductivity of the top layer: $K_1=K_2$. Unfortunately, Inouchi/Kakinuma (1992) do not provide all necessary data to recalculate the proposed constellation. The conductivity of the upper layer K_1 needs to be known (K_1 remains in the transformed set of equations (11.16)) determine the streamfunction boundary condition from the hydraulic head condition. The FAST-C(2D) model assumes K_1 to be 10^{-5} m/s. The following figures illustrate modeling results.

Parameter	Value	Parameter	Value	Parameter	Value
d_0	0.2	h_s	2.00 m	D_x, D_z	$1. \cdot 10^{-3}$
d_1	0.4	h_f	2.06 m	μ	10^{-3} kg/m/s
d_2	0.4	$\Delta \Psi$	0.003	$\Delta \rho$	25 kg/m^3

Table 11.6: Input data for the layered aquifer model (notation see Fig. 11.8)

Characteristic patterns similar to those shown in the Henry example can be observed for slightly reduced conductivity in the intermediate formation (K_0/K_1=1/5). The streamlines are steeper within this layer. The phenomenon is generally known from layered aquifer-aquitard systems: aquitards are crossed along shorter paths (see Bear/Veruijt 1987). It is also well known that this effect becomes stronger with increasing conductivity ratio. This is clearly visible in a comparison between top, center and bottom figures. For K_0/K_1=1/500 flow within the low permeable formation is almost vertical.

Another interesting feature is clearly illustrated by the bottom streamline figure. The homogeneous case boundary at the saltwater side can be divided in a lower inflow part and an upper outflow part, but the situation becomes more complicated in the layered system. Two inflow and outflow intervals can be identified in the layered system. There is a wide ranging wedge in the lower aquifer and another wedge with less extension in the upper aquifer. Each aquifer has its own saltwater intrusion. This is reasonable if there is a low permeable separating layer.

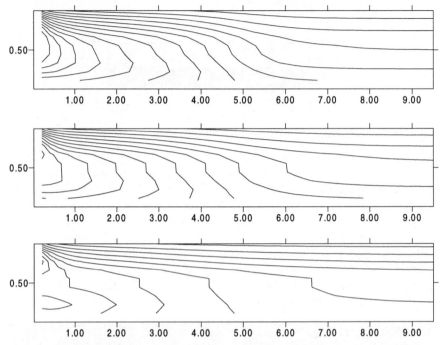

Fig. 11.9: Saltwater intrusion: streamlines for the layered aquifer system; variation of conductivity ratio between aquifers and aquitard;
top: $K_0/K_1=1/5$, center: $K_0/K_1=1/50$, bottom: $K_0/K_1=1/500$

Salt concentrations in the aquifers are quite different. An interpretation of the bottom part of Fig. 11.10 is that salinity of outflowing water in the lower aquifer is equal to the salinity of inflowing water in the upper aquifer. Salinity fronts in the aquitard are almost vertical.

There is an interesting result concerning the depth of intrusion of the salt wedge into the entire system. The saltwater front penetrates to a lesser degree, if there is an intermediate relatively impermeable layer. This result is not surprising, because the mean conductivity of all layered system cases is lower than the conductivity for the homogeneous case, when the parameters given above are used.

Qualitatively, the results coincide with those presented in the paper of Inouichi/Kakinuma (1992). A numerical comparison with the data given in the paper is not possible, as mentioned above.

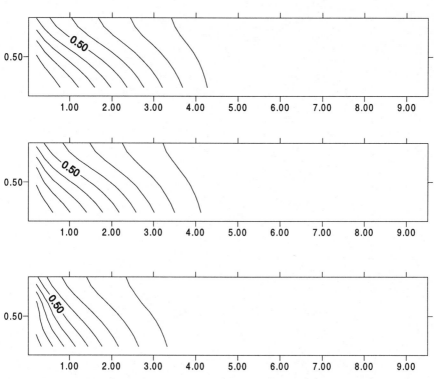

Fig. 11.10: Saltwater intrusion - isohalines for the layered aquifer system - variation of conductivity ratio between aquifers and aquitard (top: K_0/K_1=1/5, center: K_0/K_1=1/50, bottom: K_0/K_1=1/500)

12 Saltwater Upconing

12.1 Problem Description

Water with increased salt contents can often be found in the subsurface. The salt may stem from the dissolution of salt formations, such as bedded salt or salt-domes (see chapter 13). There may be 'fossil' fluids, leftovers from former geological time-periods, when parts of the today's continental surface was at the bottom of the sea. In coastal areas, saltwater from the sea penetrates into the aquifers (see chapter 11).

When a well is placed into an aquifer, where freshwater overlies a saline reservoir, some amount of salt water may be pumped after a while. This may even occur if the well is not screened in the former salt water part, i.e. if the bottom of the well lies some distance above the saline region.

One effect of pumping is a change in the original flow regime in the vicinity of the well. Water moves towards the well-screen from all directions. This leads to a lowering of the water table in the upper part of the aquifer and an upward movement of saline water in the lower part. When the saltwater wedge enters the well bottom, a mixture of saline water and freshwater is pumped. Salinity may increase gradually and may finally exceed a critical level.

The critical margin itself depends on the use, for which the water is pumped (see chapter 2: classification of saline fluids). Water with excessive salt concentrations cannot be used for drinking, for irrigation or by the industry. That is why exploitation of the aquifer should be restricted if a significant increase of salinity is observed. Pumping rates for single wells or well galleries should not be increased above a certain limit.

As described pumping in the freshwater region may cause saline waters to rise. The rise relative to the original position is highest just beneath the well bottom. The effect becomes smaller with the radial distance from the well position. Fig. 12.2 shows the situation with an interface between fresh and saline waters. a conic form of the interface surface results in three dimensions under idealized homogeneous conditions.

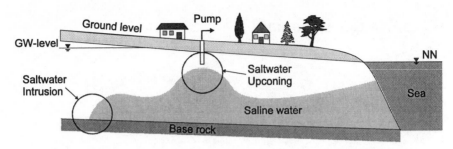

Fig. 12.1: Distinction between saltwater upconing and saltwater intrusion -
schematic view

Saltwater upconing may occur in connection with saltwater intrusion (see chapter 11). The situation is depicted in Fig. 12.1. The distinction is that intrusion is basically a horizontal movement of the saltwater wedge into the aquifer; the real flow pattern is more complicated however. In contrast, upconing is basically a vertical (upward) movement of saline fluid. If saltwater intrudes into an aquifer, upconing may occur when a pumping well is installed above the saltwater wedge.

Upconing of saltwater below pumping wells is a severe problem reported in several regions of the world. Case studies are published from Germany (Diersch e.a. 1984, Diersch/Nillert 1987), the Netherlands (Huisman 1954), Israel (Dagan/Bear 1968, Schmorak/Mercado 1969), Japan (Hosokawa e.a. 1990), Pakistan (Wang 1965, Bennett e.a. 1968) and the United States (Rubin 1983, Reilly/Goodman 1987, Reilly e.a. 1987, Motz 1992). Some of these cases concern inland aquifers and some concern coastal regions, where upconing is connected with intrusion.

12.2 Modeling Saltwater Upconing

12.2.1 Sharp Interface Approach

The first approaches for the modeling of saltwater upconing were made analogous to the problem of oil recovery treated by Muskat/Wyckoff (1935), where oil is placed above water. Most authors who treated the problem this way made three assumptions:

- the interface between both fluids (saline and fresh water) remains stable or: there is no transition zone before and during pumping from the well

- the streamlines or flowpaths (contours of streamfunction) for the pumped water do not reach the saline region or: flowpaths or streamlines (in steady state) do not cross the phase interface.

- a steady state is reached in a relatively short time after the start of pumping

Fig. 12.2 depicts the second assumption following Strack (1972).

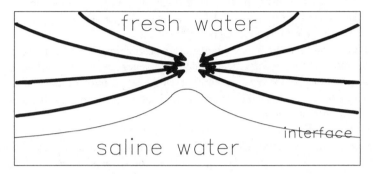

Fig. 12.2: Idealized approach to saltwater upconing

The assumptions have been basic for a series of studies: Wang (1965), Bennett e.a. (1968), Sahni (1973), Strack (1972, 1976), Motz (1992). The sharp interface assumption gives a certain highest stable position for the brine cone. The rising of the saltwater front below a well can be classified as sub-critical (no pumping of salt-water) or super-critical (pumping of salt-water). Conditions have been developed by many authors which, if considered in well design and operation, guarantee that the highest point is not exceeded. These conditions cannot be sharply formulated, but situations may be defined as being critical if they may possibly lead to saltwater pumping. An overview on the subject is given by Thiele (1983).

A more sophisticated treatment of the sharp interface problem is presented by Dagan/Bear (1968), who skip the second and third of the above listed assumptions. A second potential is defined in the saltwater region and estimations for the transient development of the upconing are derived using analytical means.

Numerical models of the problem based on the sharp interface assumption are described by Wirojanagud/Charbeneau (1985) and Reilly e.a. (1987).

12.2.2 Miscible Displacement

The classical approach implies that no salt is pumped from the saline region as long as the stability conditions mentioned above are fulfilled. There should be no increase of salinity above the low fresh water levels. But many case studies with increasing exploitation of aquifers observe a gradual increase of salinity. A zone with mixed fresh and saline fluids appears within the aquifer.

The build-up of a transition zone has already been mentioned by Huisman (1954), evaluating observations in the North-sea dunes in the vicinity of Amsterdam. Dagan/Bear (1968) write: '*Actually a transition zone, rather than an abrupt interface, exists between the two miscible liquids due mainly to hydrodynamic dispersion*'. As Ségol (1994) puts it: '*the sharp interface representation is of course a fiction; sea water and fresh groundwater are miscible fluids*'. A transition zone with brackish water will build up in the pore space. This zone may be more or less extended. Only a very small transition zone may justify the sharp interface approach.

Pumped water will be a mixture of waters originating in the fresh and the saline part of the aquifer. In practical situations, water users accept low levels of increased salinity if there is no conflict with the purpose of use. Nevertheless, mixing can only be modeled by introducing dispersion - a process that is neglected in the sharp interface approach.

The first attempt has been made by Schmorak/Mercado (1969) to describe a transition zone. A modification of the interface solutions, as derived by Dagan/Bear (1968), is proposed to account for dispersion. A boundary layer approximation for the mixing zone is used by Rubin/Pinder (1977) and Rubin (1983).

A more complex approach is to compute salinity solving a transport equation. As shown in chapter 3 the following processes can be taken into account: advection, diffusion, dispersion and sorption. There are several programs that could be used to solve the transport equation. To mention two of those: the FAST-B(2D) code, as described by Holzbecher (1995), is designed for transport in 2D flow fields; transport in 3D can be modeled using MT3D (1992).

The following presents an approach, which is capable to account for variable density effects additionally. A schematic representation of the conceptual model is given in Fig. 12.3.

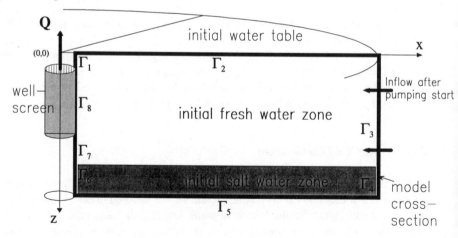

Fig. 12.3: Conceptual model and boundary types

Boundary conditions in the model at various parts Γ_i ,i=1,...8 (see Fig. 12.3) are gathered in Table 12.1. Fluid flow conditions are treated by the streamfunction formulation. The values Ψ_{high} and Ψ_{low} which have to be specified at the boundary, are derived from the pumping rate Q using the following formula:

$$\Psi_{high} - \Psi_{low} = Q \qquad (12.1)$$

The modeler has the freedom to choose one of the Ψ-values arbitrarily.

Diersch e.a. (1984) use total pressure p as flow variable and therefore need to specify boundary conditions differently. The formulation for the boundary parts Γ_i ,i=2,...8 is given in Table 12.2. The conceptual model of Diersch e.a. (1984) does not consider an unscreened well segment Γ_1.

Reilly/Goodman (1987) choose the pressure formulation as well, but use different conditions at the top and bottom boundaries (Γ_2 and Γ_5 respectively). At Γ_2, a prescribed flux condition is applied to take groundwater recharge into account. At Γ_5, a Dirichlet-condition for pressure is specified. The formulation of hydrostatic pressure at Γ_3 and Γ_4 needs to take the changing density into account where the interface meets the vertical boundary (Diersch e.a. 1984). It is even somewhat more complicated when an initial mixture zone is assumed, as was done by Reilly/Goodman (1987).

Boundary	Flow boundary condition	Transport boundary condition
Γ_1	Dirichlet type $\Psi = \Psi_{low}$	Neumann type $\partial\theta/\partial x = 0$
Γ_2	Dirichlet type $\Psi = \Psi_{low}$	Dirichlet type $\theta = 0$
Γ_3	Neumann type $\partial\Psi/\partial x = 0$	Dirichlet type $\theta = 0$
Γ_4	Neumann type $\partial\Psi/\partial x = 0$	Dirichlet type $\theta = 1$
Γ_5	Dirichlet type $\Psi = \Psi_{high}$	Neumann type $\partial\theta/\partial z = 0$
Γ_6	Dirichlet type $\Psi = \Psi_{high}$	Dirichlet type $\theta = 1$
Γ_7	Dirichlet type $\Psi = \Psi_{high}$	Neumann type $\partial\theta/\partial x = 0$
Γ_8	Neumann type $\partial\Psi/\partial x = 0$	Neumann type $\partial\theta/\partial x = 0$

Table 12.1: Overview on boundary conditions using the streamfunction formulation

The physical relevance is basically the same, but the Ψ- and p-formulation are not completely equivalent. Differences concern the boundaries Γ_3 and Γ_4, where the streamfunction formulations postulates horizontal influx. In the pressure formulation, velocity vectors at the inflow boundaries will be slightly rotated from the horizontal direction. The major difference can be found at the well screen boundary Γ_8. The p-formulation requires the normal velocity, and thus flux, to be specified in all blocks or elements along that boundary. The velocity is usually assumed to be constant, despite the fact that this is not true for real wells. The Ψ-formulation offers a better possibility to formulate the boundary condition. The jump from Ψ_{high} to Ψ_{low}, which occurs from one end of the well screen to the other, is determined by the pumping rate Q (see equation (12.1)). Outflow velocities in various depths are calculated within the modeling run.

Boundary	Pressure condition	Normal velocity
Γ_2	-	No flow: $v=0$
Γ_3	Dirichlet: hydrostatic	-
Γ_4	Dirichlet: hydrostatic	-
Γ_5	-	No flow: $v=0$
Γ_6	-	No flow: $v=0$
Γ_7	-	No flow: $v=0$
Γ_8	-	Prescribed: $v=v_{bound}$

Table 12.2: Flow boundary condition for upconing problem using pressure formulation (used by Diersch e.a. 1984)

The existence of a sharp interface is assumed as initial condition. There are two physical processes that may cause the stability of the interface in presence of the initial no-flow or low flow regime. The small diffusive flux from the saline into the fresh water region is balanced by buoyancy forces which prevent the rise of saltwater.

In the miscible displacement case with non-zero flow the stability of the interface is mainly influenced by transversal dispersivity (Wirojanagud/Charbeneau 1985, Thiele 1983). Interfaces in high velocity fields will not be stable. The start of pumping increases the velocities in the well environment up to several orders of magnitude. These conditions also favor the mixing of fresh and saline waters by transversal dispersion.

Longitudinal dispersivity plays an important role for the migration of salt components. α_L influences the concentration distribution along the streamlines connecting the saltwater region and the well screen.

Numerical models for miscible displacement have been applied in the upconing problem since the mid 80s. Diersch e.a. (1984), Reilly/Goodman (1987) and Diersch/Nillert (1987) use the FE (finite element) method and the pressure formulation. Holzbecher (1995) and Holzbecher/Heinl (1995) apply the FD (finite difference) method and the streamfunction formulation. In the FE-simulations take into account a velocity dependent dispersivity, while mentioned FD-simulations work with a constant diffusivity. Holzbecher/Heinl (1995) consider numerical dispersivity.

Reilly/Goodman (1987) construct four cases to compare the output of a miscible flow/variable density model with the predictions of the sharp interface location from a former study (Bennett e.a. 1968). All defined cases describe subcritical situations. Calculations are made with the SUTRA-code (Voss 1984) using the steady state option.

Reference	Diersch e.a.	Reilly/ Good- man 1	Reilly/ Good- man 2	Reilly/ Good- man 3	Reilly/ Good- man 4	Holz- becher	Holz- becher/ Heinl	Diersch / Nillert
Q	100 m³/h	56 kg/s	30 kg/s	21.3 kg/s	10.65 kg/s	100 m³/h	100 m³/h	50/75 m³/h
φ	.36	.2	.2	.2	.2	.3	.3	.3
k_r	$1.13\ 10^{-4}$ m^2	2.75 $10^{-10}ft^2$ 2.5 $10^{-11}\ ft^2$	2.75 $10^{-10}ft^2$ $2.5\ 10^{-11}$ ft^2	$1.55\ 10^{-9}$ ft^2 1.44 $10^{-10}ft^2$	$1.55\ 10^{-9}$ ft^2 1.44 $10^{-10}ft^2$	$6.\ 10^{-4}$ m^2	$6.\ 10^{-4}$ m^2	$1.4\text{-}6$ $10^{-4}\ m^2$
k_z	$1.13\ 10^{-4}$ m^2	2.75 $10^{-10}ft^2$ $1.0\ 10^{-11}$ ft^2	2.75 $10^{-10}ft^2$ $1.0\ 10^{-11}$ ft^2	$1.55\ 10^{-9}$ ft^2 1.44 $10^{-10}ft^2$	$1.55\ 10^{-9}$ ft^2 1.44 $10^{-10}ft^2$	$6.\ 10^{-5}$ m^2	$6.\ 10^{-5}$ m^2	$1.4\text{-}6$ $10^{-4}\ m^2$
α_L	1m / 3m	1 m	1 m	1 m	1 m	-	-	.75 m
α_T	.1m/ .3m	.5 m	.5 m	.5 m	.5 m	-	.125-.5m	.0375 m
L	75 m	3280 ft	3280 ft	2980 ft (=908 m)	2980 ft (=908 m)	94.3 m	94.3 m	
$\Delta\rho\ (\%)$	.1, .3, .5	2.5	2.5	2.5	2.5	.6	.6	.46
S	0m / 5m	38.75 ft	38.75 ft	15 ft	15 ft	20 m	20 m	
transit[1]	no	yes	yes	yes	yes	yes	yes	
steady / transient	s,t	s	s	s	s	t	t	t
super / subcrit.	super / sub	sub	sub	sub	sub	super	super	super

Table 12.3: Overview of saltwater upconing model parameters as published by Diersch e.a. (1984), Reilly/Goodman (1987), Holzbecher (1995), Holzbecher/Heinl (1995), Diersch/Nillert (1987)

Reilly/Goodman (1987) find that the 50% isochlor contour from the numerical model and the sharp interface are very close. The transition zone seems to be

---
[1] Initial transition zone

asymmetric around the sharp interface. Salinity levels in pumped water reach 10% salinity at the aquifer bottom, while the sharp interface approach predicts no increase. As expected, transverse dispersivity has a significant influence on the spreading of the mixture zone, but the effect of longitudinal dispersivity on steady state salinity patterns is marginal.

Diersch e.a. (1984) compare results obtained using the sharp interface assumption and the results of methods assuming miscible displacement. They find a good agreement with the theoretical findings of Dagan/Bear (1968) in the subcritical range, i.e. when upconing is small. As expected, the agreement is better for small dispersivity. Numerical methods in the critical and supercritical case give salinity levels much higher than predicted by Dagan/Bear (1968).

12.2.3 Variable Density Effects

If miscible displacement is taken into account, the coupling of water flow and salt transport has to be considered as well. Salt concentrations may influence the fluid density and viscosity. The flow pattern itself may be altered by changing salinity. This is the typical situation for density-driven flow. If the salinity in the lower part of the system is relatively low, and density of saltwater is only slightly greater than fresh water density, variable density effects can surely be neglected.

The horizontal interface is influenced by two counteracting processes. The upward-directed processes of diffusion and transversal dispersion tend to dissolve the interface to set up a mixture zone of a certain width. The downward-directed process of gravity (the counterpart to buoyancy) within the pores tends to separate the light fresh water at the top from the heavy brackish water at the bottom (for high saline brine only).

The above mentioned studies all take density effects into account: Diersch e.a. (1984), Reilly/Goodman (1987), Diersch/Nillert (1987), Holzbecher (1995) and Holzbecher/Heinl (1995). The density difference has been quite low in most case studies and variable density effects have not been recognized. Holzbecher/Heinl (1995) report that the required accuracy for the solution in each timestep was reached after the first Picard iteration. This is a clear indication that variable density effects are marginal. The model of Diersch e.a. (1984) in contrast required 1 to 6 Picard iterations in the transient case and 4 to 13 iterations for the steady state to converge. Unfortunately, none of the mentioned studies made a calculation with a hypothetical tracer in order to separate dispersion and variable density effects.

An 'overshooting' effect is observed by Diersch e.a. (1984) with increasing density difference. Salinity in pumped water rises to a maximum and later falls to a final steady value. The effect does not appear if density differences are low and if a velocity-independent diffusivity is used for modeling. Overshooting is caused by the interaction between dispersion and density effects. Diersch e.a. (1984) state that *'overshooting is primarily of a theoretical nature'*, but the point is further discussed by Thiele/Diersch (1986).

Variable density effects have been mostly neglected in scientific literature, despite the availability of numerical codes for this type of study. Saltwater upconing in the critical and supercritical range is probably underestimated when the sharp interface assumption is used (Diersch e.a. 1984). Pure dispersion models - with constant density - certainly overestimate upconing. It is precisely the effect of variable density to narrow the transition zone and prevent the overestimation. It would be interesting to get a quantification of this effect from numerical calculations. If it turns out that density effects are marginal under realistic conditions, usual (constant density) transport codes could be used to treat saltwater upconing.

12.3 Case Study

As an example the case of a shallow aquifer is treated. A fresh water reservoir of approximately 30m depth overlies a region of saline water. The aquifer in question is located in Guddu (Pakistan). A gallery of pumping wells was to be established in shallow fresh water to provide cooling water for a power plant. One well was selected for simulation purposes. Salinity in this well rose to an unacceptable level within a five month period. During that time electrical conductivity was measured regularly.

Fig. 12.3 depicts the conceptual model of the well environment. Table 12.4 provides the lengths of the various boundary segments which are shown in Fig. 12.3. Table 12.5 lists the physical characteristics.

Boundary	Γ_1	Γ_2	Γ_3	Γ_4	Γ_5	Γ_6	Γ_7	Γ_8
Length [m]	6	94.3	30	20	94.3	20	3	21

Table 12.4: Geometrical parameters for Guddu case study

The FAST-C(2D) code was applied to model the observed situation with the conceptual model given above in Fig. 12.3. Mixing of waters with different salt contents is taken into account, as well as variable density effects. Holzbecher (1995) published a first approach for the field situation and showed that variable density effects are negligible. Some details in the results indicate that grid spacings in that model were too wide. The grid has been refined twice in the vertical direction. Most other settings turned out to be satisfactory in the previous simulation and have been kept in a second publication (Holzbecher/Heinl 1995).

Holzbecher/Heinl (1995) finally used an irregular rectangular grid of 10×98 blocks for the numerical simulation[2]. Blocklengths in vertical direction vary from 0.25 m to 1 m. In horizontal direction grid spacing increases linearly with distance from the well. The smallest block at the well edge extends to 0.23 m; the outermost block measures 38.6 m.

Diffusivity in natural systems is usually in the range of 10^{-9} m^2/s. This is low in comparison to hydrodynamic dispersion. The model set-up by Holzbecher/Heinl (1995) does not include dispersion, but does have numerical dispersion. The upwind scheme was used for the spatial discretization of 1st order terms. Truncation error analysis yields a formula for numerical dispersion in z-direction (see equation (4.7)):

$$D^{vertical} = \frac{\Delta z}{2} v_z \qquad (12.2)$$

Obviously this represents a dispersion length of $\Delta z/2$. The grid data in the presented model indicates that the numerical dispersivity in z-direction lies between 0.125 m and 0.5 m (see equation (12.2)) - depending on the vertical grid spacing. Within the transition zone numerical dispersivity is 0.125 m, i.e. near to values used in other field models (Diersch e.a. 1984, Reilly/Goodman 1987, Momii e.a. 1989, Hosokawa e.a. 1990). This value is the vertical dispersion of the discrete model.

Parameter	Value	Parameter	Value
Hydraulic conductivity (horizontal)	$6 \cdot 10^{-4}$ m / s	Well radius	0.35 m
Hydraulic conductivity (vertical)	$6 \cdot 10^{-5}$ m / s	Pumping rate	100 m^3 / h
Porosity	0.3	Constant dynamic viscosity	10^{-3} kg / m / s
Fresh water density	1000 kg / m^3	Streamfunction at aquifer top	0.02778 m^3/s (=100/3600)
Salt water density	1006 kg / m^3	Streamfunction at aquifer bottom	0. m^3/s

Table 12.5: Physical characteristics for Guddu case study

---

[2] The input data for the model can be found on CD-ROM in the file *Upconing.WTX*

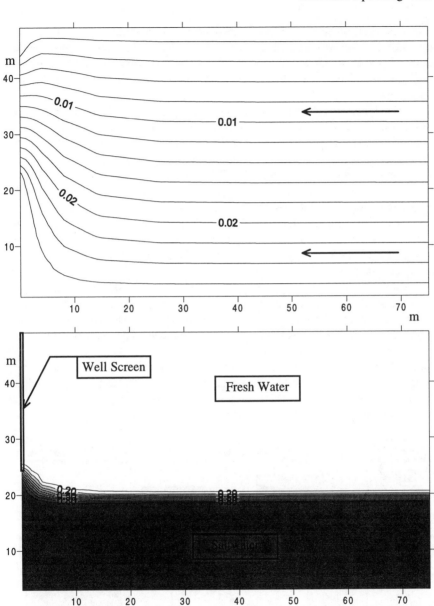

Fig. 12.4: Results from the saltwater upconing example modeled by FAST-C(2D)
top: streamlines; below: isohalines for levels 0.1(0.1)0.9 of saltwater
concentration one month after start of pumping

Because the dominant flow direction is horizontal, transversal dispersion mainly
acts in vertical direction. The simulated numerical dispersivity is approximately
half of the grid-spacing in vertical direction (see equation 12.2).

An (almost) steady state was reached after a long term simulation run. The intermediate state was reached after one month. Fig. 12.4 depicts the intermediate state and shows streamlines and lines of constant concentration. Table 12.5 gives selected input data. The grid has 10 blocks in x-direction and 49 blocks in z-direction.

Selected timesteps were longer than allowed by the Courant criterion. But results were affected only marginally. The reason is probably that high velocities can be observed only in the vicinity of the well, which is a small part of the entire system. Further tests concerning the influence of the numerical parameters would surely be necessary in an expert study but can be omitted here.

The transient development of the salinity levels of pumped water is given in Fig 12.5. The measured data shown in the graphics were not obtained under a controlled experiment. Pumping was stopped several times leading to short time decreases of observed salinity. The pumping rate was not constant and thus the measurements are used only to make sure that timescale and salinity scale are in a reasonable range.

It can be noted that salinity remains on a low level for quite some time until the increase starts. The curve reflects that by switching from a concave to a convex shape. The observed data show the same behavior. The output of the numerical model gives the physical explanation for the mentioned delay in salinity increase. In the beginning the salt, which reaches the well, is mainly transported by longitudinal dispersion along the fast streamlines connecting inflow boundary and well screen. Salinity in the low flow region underneath the well reaches a higher level only after some time. Once the upconing has reached the well bottom, salt concentrations in pumped waters start to increase at a higher pace.

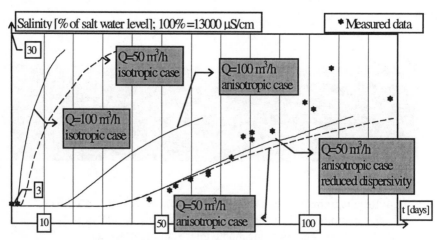

Fig. 12.5: Transient salinity increase for various situations

Breakthrough curves are shown for different (constant) pumping rates under the assumption of an isotropic and an anisotropic porous medium. The characteristic behavior of the data (delayed salinity increase) can be explained only by the model that considers anisotropy.

Reduced (numerical) dispersivity in a model with a 10×196 grid show the influence of dispersion. Obviously the steep gradient of the concentrations front increases when dispersivity decreases and thus the coincidence with measured data becomes better.

13 Flow across a Salt-Dome

13.1 Salt Formations and Scenarios

Groundwater flow in geologic formations overlying a salt-dome has become of special interest since salt formations have been considered for the disposal of hazardous waste. There are plans in Germany to use the Gorleben salt-dome for the storage of high level radioactive waste (HLW) from nuclear reactors. The USA Waste Isolation Pilot Plant (WIPP) is located in a bedded salt formation and is intended for transuranic waste. Denmark and the Netherlands have explored the salt option too. A recent survey lists Germany, Spain, the Netherlands, the USA and former USSR states as nations planning to use salt formations for disposal of HLW (McCombie 1997).

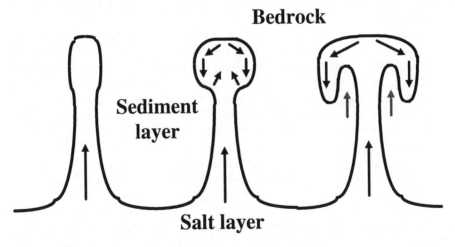

Fig. 13.1: Schematic illustration of three types of salt-domes
(see Jackson/Talbot 1986, Talbot/Jackson 1987)

A salt-dome is a geological formation in the subsurface in which salt rises through sediment layers towards the surface of the earth. The origin of the formation can be imagined as a salt layer gradually covered by sediments. Consolidation of the covering layers over geological timeperiods leads to an unstable situation. An initially stable series of layers tends to become unstable when material of higher density overlies material of lower density. The incompressible salt with high plasticity starts to rise through the sediments at certain locations forming a typical dome. Fig. 13.1 illustrates that the flow pattern in the upper part of the salt dome may be quite complex.

Salt layers or domes are viewed as good host formations for heat producing waste because they have

- high plasticity: the flow of salt closes fractures in short times
- high thermal conductivity

Some authors doubt about that because of the

- impurity of the salt
- salt-dome diapirism - leading to complex formations and rock structures at the salt-dome boundaries
- solubility of salt - leading to a fast dissolution of salt in contact with water
- corrosion effects of salt

Safety analyses for waste deposits of this type are often based on scenarios. One scenario assumes that all barriers inside the salt formation fail to work and that a huge amount of toxic and radioactive material enters the overlying aquifer and migrates with the flow of water. Radionuclides may reach the earth's surface across permeable channels and cause damage to human beings and the environment.

There are a lot of uncertainties in this scenario and the uncertainties are accumulated because the timeperiod of interest is determined by the halflife of the material with slowest decay. The legislation of different nations set different requirements for the safety of HLW repositories. The US-EPA (Environmental Protection Agency), for example, requires that for a time period of 1000 years radionuclide concentration in groundwater remains below a certain limit. Moreover, cumulative release of radionuclides to the environment from all significant processes should be limited for a period of 10000 years (Brown/Lemons 1991). The 10000 year period is quite arbitrary and it is reasonable to consider even longer timeperiods (Kastenberg/Gratton 1997).

The long periods of time involved are not the only factor that makes modeling seem the only way to predict the possible migration of hazardous waste in the subsurface. Another factor that speaks for modeling is that during the migration there is a complex interaction of processes. Transport simulations for radionuclides take into account advection, diffusion, dispersion, sorption and decay (Bütow e.a. 1985). The basis for such simulation is knowledge of the flow field in the aquifer. A schematic view of subsurface conditions is given in Fig. 13.2.

The fluid in the vicinity of the salt-dome is highly saline. It is hotter than the overlying layers because of the geothermal gradient. Temperatures in a salt-dome containing heat producing waste may be even higher. The penetration of radioactive, heat emitting materials may be of additional influence. All these processes cause the fluid properties, density and viscosity, to change spatially and temporally.

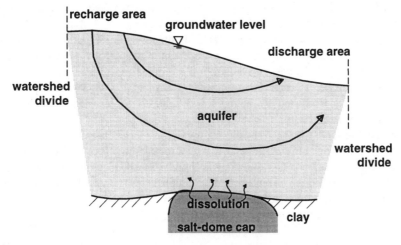

Fig. 13.2: Schematic view of salt-dome top and overlying aquifer

This is a typical situation of density driven flow and has been described in chapter 1. Density gradients may be the reason for flow patterns that differ from the common potential regime in constant property fluids. The flow pattern on the other side may cause density differences. Flow and salt transport interact with each other. The distribution of radionuclides in a flow field caused by density changes is different from the distribution in another flow field without density changes.

A test-case from the international verification and validation project HYDROCOIN will be discussed in detail. The model region in the test-case concerns density driven flow in the aquifer overlying the upper edge of the salt-dome. Leijnse (1985a) proposed and discussed a different but very similar hypothetical conceptual model. In general the geological situation may be much more complicated. Evans/Nunn (1989) and Evans e.a. (1991) point out that vertical upward fluxes have been observed along the sides of some salt formations in Louisiana - a situation clearly too complicated to be explained by the idealized model treated below.

13.2 HYDROCOIN Test-Case

The international HYDROCOIN workshop was initiated and organized by the Swedish Nuclear Power Inspectorate (SKI). Its aim was: *'to obtain improved knowledge of the various strategies for groundwater flow modelling for the safety assessment of final repositories for nuclear waste. To this end, calculations are made with different mathematical models used by a number of organisations'* (HYDROCOIN 1988). The first step was to define and agree about test-cases to be treated by participating teams.

Some test-cases of the project tackled situations with saline or thermal gradients. As a thermal case the Elder experiment of chapter 9 was treated in level 2 case 2. A saline case was derived from the problem of disposal in a salt-dome and appeared in level 1, case 5, and in level 3, case 4[1]. It was originally proposed by Bütow (1984) assisted by Heredia and selected for sensitivity analysis by Bütow/Holzbecher (1986).

The background for the test-case set-up is the Gorleben salt-dome in Lower Saxony in Germany. The salt-dome is foreseen to hold high level radioactive waste produced in Germany. The situation is highly idealized. The geological and geometrical conditions are intentionally simplified in order to focus on the mathematical methodology of density driven flow. The model region is a vertical cross-section through the aquifer as shown in Fig. 13.3.

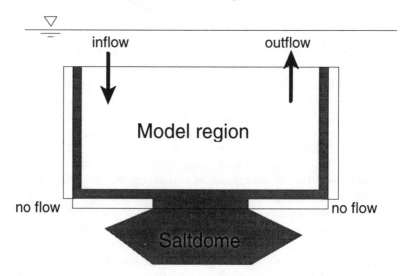

Fig. 13.3: Schematic view of the HYDROCOIN test-case; salt-dome problem

---

[1] HYDROCOIN is divided into three levels: 1, code verification; 2, model validation; 3, sensitivity and uncertainty analysis

The model region is closed at the bottom and at both vertical boundaries. A linear changing pressure field is prescribed at the top, i.e. there is a constant horizontal velocity component. The no-flow condition is postulated for salt at closed boundaries, i.e. the normal derivative of concentration is zero. At the bottom of the model region the upper surface of the salt-dome is situated. The constellation is expressed by a special boundary condition, i.e. a condition of Dirichlet-type with predefined salt concentration. A schematic view of the boundary conditions is given in Fig. 13.3. The value of salinity is not specified explicitly in the HYDROCOIN test-case. Table 13.1 provides an overview of other parameters.

Parameter	Value
permeability	10^{-12} m^2
porosity	0.2
viscosity	$1.0 \ 10^{-3}$ Pa s
fresh water density	1000 kg/m^3
saline water density (max.)	1200 kg/m^3
longitudinal dispersivity	20 m
transversal dispersivity	2 m
pressure difference at top boundary	10^5 Pa
model depth	300 m
model length	900 m

Table 13.1: Input data for the test-case HYDROCOIN level 1 case 5

The final report (HYDROCOIN 1988) contains some remarks on the test-case definition given by the participants of the HYDROCOIN project. No molecular diffusivity is specified. This is unrealistic and caused numerical problems. The point is treated in more detail below.

Boundary conditions at the top corners of the model are contradictory. The vertical sides condition is a zero derivative, while the top assumes a linear change of pressure, i.e. a nonzero velocity. The contradiction can be found in other concepts as well such as the Tóth case (1962, see chapter 10). For the Tóth case there are analytical solutions and good numerical approximations - despite the contradiction at the top corners.

To overcome the mentioned contradiction Leijnse (1985b) proposed the introduction of a small impermeable boundary on both sides of the top model edge. Guvanasen (1985) recommended more clearly defined recharge and discharge boundaries. He suggested the division of the top boundary in three parts: a recharge, a no-flow and a discharge part. In the recharge part Guvanasen (1985) proposed to specify a third type boundary condition. These recommendations did not lead to a change of the test case definition.

Lever/Jackson (1985) discuss the functional form of density dependency on salt concentrations. They show by basic analysis that the functional form is different for salt volume fractions or mass fractions. For the volume fraction a linear relation between density and salinity results, while for mass fractions c_m the following form is obtained:

$$\rho^{-1} = \rho_0^{-1} + c_m\left[(\rho_0 + \Delta\rho)^{-1} - \rho_0^{-1}\right]$$

(13.1)

It turns out that formulation (13.1) differs only marginally from linear relationship when parameters are determined for relevant salinity intervals. Thus major differences in the flow field resulting from different density functions can not be expected.

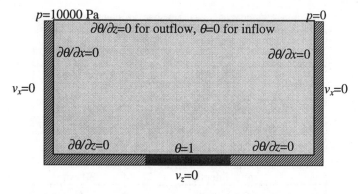

Fig. 13.4: Schematic view of boundary conditions in the original case formulation

Holzbecher (1991a) shows that an analytical solution exists for the system with the boundary condition $\partial\theta/\partial z=0$ on the top. The asymptotic situation without any density differences, where high saline water has filled the entire pore space, is a mathematical solution for differential equation and boundary conditions. The corresponding head field is given by the Tóth solution (10.1) with appropriate values for H, L, h_{min} and h_{max}. Of course this is not a realistic situation.

13.3   Modeling the HYDROCOIN Test-Case

At the time that the HYDROCOIN test problem was defined people were not aware that salt gradients could affect the flow regime in the subsurface. When numerical results for the test-case from participating groups were presented first[2], a broad discussion emerged if these patterns could be real and are not only a result of numerics. Doubts vanished with several teams showing up with *eddies* in their

---

[2] For me the first presentation of these results at a meeting (1985 in Paris) became the starting point for an intense and long lasting interest in *density-driven porous media flow*.

results. The third phase of HYDROCOIN tackled the salt-dome problem for parameter variations and sensitivity analysis (level 3 case 4).

The HYDROCOIN level 1 final paper (1988) reports six different model approaches from five groups. The following codes have been used as modeling tools: SWIFT, NAMMU, CFEST and METROPOL. Results from Voss using SUTRA (1984) are also mentioned.

Three models of the HYDROCOIN participants are set up using FD (finite differences) and three models use FE (finite elements). The number of elements or blocks ranges from 396 up to 2731. Numerical problems are mentioned by all groups. One team switched to a variation of the original case with zero dispersivity; another team used a Dirichlet boundary condition for salt (fresh water inflow) at a specified part of the top boundary (HYDROCOIN 1988).

An obstacle is zero diffusivity that causes problems in regions with very low velocities. Difficulties have been resolved by a parameter stepping technique for diffusivity using the Newton-Raphson scheme. A detailed treatment of the difficulties and the solution is given by Herbert e.a. (1988).

Teams using Finite Element codes report difficulties in the velocity calculation. Difficulties could be solved using 'mixed interpolation' for pressures obtained with quadratic elements.

A comparison of outputs obtained by the groups shows significant differences. The results of three teams are very close and physically consistent. The final report on HYDROCOIN level 1 states: *'There was general agreement that the flow consisted of the main cell driven by the imposed pressure gradient at the top surface, a small weak convection cell in the bottom left corner of the domain driven by the salt source and a weak convection cell in the bottom right corner driven by the salt source. The left hand convection cell is dragged up to the right by the main flow cell. This consensus was achieved only after considerable discussion. In the early workshop the existence of the convection cells and their numbers were vigorously debated'* (HYDROCOIN 1988). Results on concentration distributions look less different than results for flow patterns.

Herbert e.a. (1988) published their HYDROCOIN results using NAMMU code in more detail. SWIFT output is presented and discussed by Bütow/Holzbecher (1987). Bouhlila (1990) mentions difficulties with the velocity formulation in Finite Elements which she tried to overcome using a parameter stepping technique for the value of gravity g. First results with FAST-C(2D) were presented in Holzbecher (1991a), and an extension of that model is given in part 13.4 below. Leijnse (1992) discusses the influence of various strategies to resolve the nonlinearities and finds a severe dependency. Oldenburg/Pruess (1995), Johns/Rivera (1996), Oldenburg e.a. (1996), Konikow e.a. (1997) and Kolditz e.a. (1998) reported on modeling results more recently.

Diffusivity has become the most discussed parameter in publications on the salt-dome case. Herbert e.a. (1988) applied the NAMMU code and used a parameter stepping approach, reducing diffusivity from 10^{-6} to 10^{-8} m^2/s (note that in the original definition a zero value is proposed). Oldenburg/Pruess (1995) used the

TOUGH2 code and obtained similar recirculating flow patterns for $D>10^{-7}$ m^2/s. A further reduction of D led them to '*swept-forward*' solutions, in which the eddies disappear. Because there is no advective transport from the salt-dome edge to the upstream (left side in figures) boundary, significant salt concentrations can only be observed downstream (right).

Oldenburg/Pruess (1995) believe their results are better, because they have been obtained using a direct approach and no parameter stepping. Another argument for their solutions is that the accuracy of the solution is calculated by the residual and not by change. In spite of that Johns/Rivera (1996) doubt the 'swept-forward' solution. They recalculated the case with NAMMU (1985) and a new parameter stepping and obtained eddies even for the small diffusivity of $5\cdot10^{-8}$ m^2/s. Their results are confirmed by the most recent publication from Kolditz e.a. (1998). The authors of that study used both ROCKFLOW and FEFLOW codes and found circulating flow for zero, low and high values of diffusivity and dispersivity. A discussion of these points using FAST-C(2D) code will be published at another place.

Reference	D [m^2/s]	α_L [m]	α_T [m]
HYDROCOIN level 1 case 5	0	20	2
HYDROCOIN level 3 case 4	$1\cdot10^{-8}$	20	2
Herbert e.a. (1988)	$1\cdot10^{-6}/ 1\cdot10^{-7}/ 1\cdot10^{-8}$	0	0
''	$1\cdot10^{-8}/ 2.78\cdot10^{-9}$	20	2
Holzbecher (1991)	$1\cdot10^{-6}$	0	0
Leijnse (1992)	0	20	2
Oldenburg e.a. (1995)	$1\cdot10^{-6} - 5\cdot10^{-8}$	0	0
''	0	20	2
Johns/Rivera (1996)	$5\cdot10^{-8}$	0	0
Oldenburg e.a. (1996)	$5\cdot10^{-7}/ 5\cdot10^{-8}$	0	0
Konikow e.a. (1997)	0	20	2
Kolditz e.a. (1998)	$5\cdot10^{-6}/ 5\cdot10^{-7}/ 5\cdot10^{-8}$	0	0
''	$5\cdot10^{-8}/ 1.39\cdot10^{-8}$	20	2
''	0	20	2

Table 13.2: Overview on diffusivity / dispersivity parameter values used in various studies on the HYDROCOIN salt-dome test-case

The original definition of the test case specified the zero diffusivity with the argument that molecular diffusion is small in relation to dispersion i.e. macrodispersion. This is certainly true along the main flowpaths from the salt-dome edge towards the water table. But, in those regions of the model where velocities are extremely low diffusion becomes the relevant mixing process, and a zero diffusivity is not realistic. The argument, formulated by Herbert e.a. (1988) and others, has been adopted by Bütow/Holzbecher (1986) in the definition of a new reference case with $D=10^{-8}$ m^2/s.

Table 13.2 provides an overview on values for diffusivity and dispersivities which have been used in various studies so far.

It has to be considered that the simulation of zero diffusivity and low velocity situations by numerical codes is limited. Thus the specification of zero diffusivity puts additional stress on the numerical solution, which is not justified by an analogue in real systems. A relevant check for numerical stability is the grid Péclet criterion for diffusivity D or dispersivity α_L:

$$\frac{u\Delta x}{D} \leq 2 \qquad \text{or} \qquad \frac{\Delta x}{\alpha_L} \leq 2 \qquad (13.2)$$

The MOC technique (method of characteristics) allows the numerical simulation of zero diffusivity. Konikow e.a. (1997) use the MOCDENSE code and obtain eddy solutions. Concentration contours for their model are similar to those obtained by the HYDROCOIN participants. As an additional variation they calculate the unrealistic parameter choice: $D=\alpha_L=\alpha_T=0$, i.e. a zero diffusivity and zero dispersivity case. This variation is not possible for standard Finite Difference or Finite Element codes. MOCDENSE is based on the method of characteristics and allows this variation (Sanford/Konikow 1985). From some inconsistencies in the result Konikow e.a. (1997) start to discuss the boundary condition at the salt-dome edge. They show that the introduction of a boundary layer at the top of the salt-dome leads to 'swept-forward' solutions as they have been obtained by Oldenburg/Pruess (1995).

13.4 FAST-C(2D) Model

A model of the salt-dome problem calculated using the FAST-C(2D) software is presented.[3] Boundary conditions and model region are taken as shown in Fig. 13.5. Note that inflow and outflow parts of the upper boundary are automatically determined by the program. Input data can mostly be taken directly from Table 13.1. The velocity on the upper boundary needs to be transformed in a streamfunction boundary condition. The streamfunction derivative in the vertical direction is equal to the horizontal velocity component:

---

[3] The input data for the model can be found on CD-ROM in the file *SaltDome.WTX*

$$\frac{\partial \Psi}{\partial z} = -v_x = \frac{k}{\mu}\frac{\Delta p}{L} = \frac{1}{9}10^{-6}\frac{m}{s} \qquad (13.2)$$

The transformation into a system with dimensionless coordinates, as formulated in chapter 3 (equation (3.38)), leads to the following values for the dimensionless streamfunction derivative and the Rayleigh number:

$$\frac{\partial \Psi}{\partial z} = 33.3 \qquad \qquad Ra = 588.3 \qquad (13.3)$$

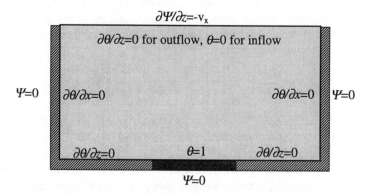

Fig. 13.5: Schematic view of boundary conditions in streamfunction formulation

Parameter / Method	Value / Type
Number of blocks in horizontal direction	36
Number of blocks in vertical direction	20
Accuracy	10^{-5}
Maximum number of Picard iterations	25
Maximum number for CG-solver	100
Solution mode	single precision
Criterion type	residual, max-norm
Timelevel weighting	Crank-Nicolson
Flow solver type	Preconditioned CG
Transport solver type	Bi-CG
1st order term discretization scheme	upwind (BIS)
Correction of numerical dispersion	none
Automatic timestepping	Courant criterion (χ=1)

Table 13.3: Overview on numerical parameters and methods for modeling the salt-dome case with FAST-C(2D)

System extensions are: $L=3$ and $H=1$. In a first approach dispersivity is set to zero in the reference case. A constant diffusivity of 10^{-6} m²/s is chosen for the whole model region. A case with identical choice of parameters was first calculated by Herbert e.a. (1988). The reference case here is given using the numerical parameters and approaches as listed in Table 13.3.

The steady state option does not converge; for that reason a transient modeling approach where the flow pattern develops with time is chosen. The starting point for the solution is the case with zero arrays for Ψ and θ. This case makes no sense physically because it does not fulfill the boundary conditions. But this is not relevant here. Any starting point could be taken that leads the transient procedure to a steady state. The time-independent solution is reached when the variables Ψ and θ show only marginal changes over a long timeperiod (here: $t_{max}=0.009$).

Fig. 13.6 shows the steady state results. The upper part shows streamlines for equidistant spacing (0.5 in the dimensionless system). Maximum value of Ψ is $\Psi_{max}=7.18$. Minimum value is taken in the center of the eddy in the left lower quarter of the model and it is $\Psi_{min}=-2.54$. There are two eddies in the bottom part of the region. Both eddies take fluid along the impermeable surface from the edges of the salt-dome towards the vertical boundaries. Reaching those boundaries the flow direction turns back in the opposite direction.

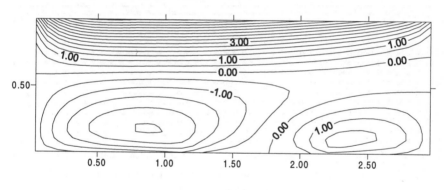

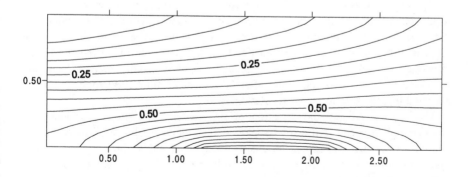

Fig. 13.6: Salinity and streamfunction contours for the salt-dome case with diffusivity $D=10^{-6}$ m^2/s, $\alpha_L=\alpha_T=0$.

The upper part of the region sees a half-eddy which is driven by the pressure gradient at the top surface. In other words: the meteoric water flow in the aquifer drives the water in the upper part of the aquifer. The emergence of the two eddies near the bottom of the model region is due to density differences. These eddies would not appear in a constant density system.

The discretization selected in the reference case is not fine enough: grid-Péclet numbers reach a maximum of 2.43 in horizontal direction. The vertical direction maximum is 1.91. Both values are below the critical margin of the grid-Péclet criterion (see chapter 4). The results coincide very well with those presented by Herbert e.a. (1988) and Oldenburg/Pruess (1995) for the same values of diffusivity and dispersivities. Holzbecher (1991a) uses a different time transformation and a different Rayleigh number; that explains the output differences.

For a sensitivity analysis of the salt-dome test-case Bütow/Holzbecher (1986) suggest varying of diffusivity/dispersivity, permeability, porosity and viscosity. Anisotropic and inhomogeneous cases and variations of boundary conditions are also proposed.

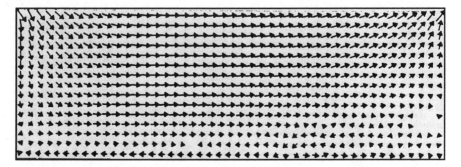

Fig. 13.7: Variation of the salt-dome problem: the anisotropic case; calculated with the SWIFT code

Fig. 13.7 depicts the flow field of the anisotropic case using an arrow field. The results have been obtained using the code SWIFT (1982) in a transient simulation run starting from a hypothetical zero salinity situation. The permeability in the vertical direction k_z is 10 times smaller than the permeability in the horizontal direction. k_x is kept from the reference case. Only small changes from the portrayed state could be observed after 500 years of simulation time. It can be expected that the solution is nearly stationary.

The arrows indicate the direction of the velocity and the vector length is proportional to speed. No arrow is plotted if velocity is almost zero. As in the reference case two eddies can be clearly recognized above the lower boundary. The extension in vertical direction of the rolls into the overlying aquifer is smaller in the anisotropic porous medium.

As noted above the diffusivity value was changed in the test-case definition of HYDROCOIN level 3 (see Table 13.2). Results for the new reference case, calculated with FAST-C(2D) are depicted in Fig. 13.8.

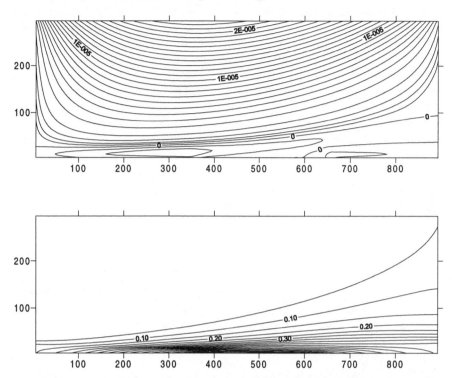

Fig. 13.8: Salinity and streamfunction contours calculated with FAST-C(2D) for HYDROCOIN level 3 case 4 with for the salt-dome case with diffusivity $D=10^{-8}$ m^2/s, $\alpha_L=20$ m, $\alpha_T=2$ m.

A series of model runs using FAST-C(2D) was performed to study the effect of the vertical boundaries. Holzbecher (1991b) points out that the horizontal extension of the region is not realistic. A distance of at least several kilometers can mostly be found between two water divides of the specified depth of 300m. A typical extension of drainage areas is discussed by Zheng/Bennett (1995). They use a more complex boundary condition first introduced by Tóth (1963).

Holzbecher (1991b) published results for systems with larger extension in x-direction, i.e. higher values for L. The length of the salt-dome cap (300m) and all other parameters from the reference case have been kept in the studies - as noted in equation (13.3) and in Table 13.1. It turned out that the flow pattern above the salt-dome remains the same. There is almost no influence from the limiting vertical water divides on the flow pattern above the salt-dome.

Another series of model runs has been performed to check the sensitivity of results concerning the dimensionless parameters Ra and Co. The parameter combination Co has been introduced in chapter 10 and has been used successfully to characterize the penetration of saltwater intrusion in chapter 11.

As outlined in chapter 10 Co relates convective versus potential flow. In this chapter it is used in a similar manner. The intention is to study the influence of Ra and Co on the flow pattern. Particularly the steady state convection roll in the lower left part of the model cross-section is of interest. It has been selected as a characteristic for the system response to changes of input parameters.

The general outlook of the streamfunction as a surface in 3D-space is presented in Fig. 13.9. Ψ takes its minimum in the center of the convection current above the left part of the bottom edge. Ψ_{min} in the following refers to the minimum of the streamfunction on the grid, which has been used for the numerical model. No interpolating post-processor is used.

The solutions for Ψ may show one or two local maxima. One stems from the potential flow and is located somewhere in the upper block line - it is noted as Ψ_{max1}. Another maximum (Ψ_{max2}) may be found in the lower right part of the model area. Fig. 13.9 shows such a situation.

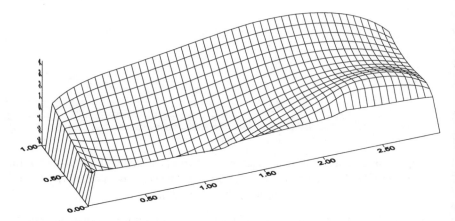

Fig. 13.9: View of streamfunction surface (here for Ra=400 and Co=40)

In the sensitivity analysis Ra has been chosen in the interval [100,600]. Co values lay between 3.333 and 40. For lower values of Co there is no visible change in the flow patterns. Thin long eddies at the bottom can hardly be recognized.

Co	3.333	6.667	13.333	20	30	40
Block-index	39/40	37/38	35/36	34	32/33	32/33
Distance [m]	7.5	22.5	37.5	48.75	60.0	60.0

Table 13.4: Dependence of convection eddy center on Co

The model uses an equidistant grid with 72×40 blocks. All other parameters are chosen as listed in Table 13.3. The discretization method is BIS, i.e. there is significant numerical dispersion in some model runs.

The most obvious result is that the distance, in which Ψ_{min} can be found from the lower boundary, is mainly determined by the parameter Co. In other words: the center of the convection eddy in the lower left part is almost independent from the Rayleigh number. This holds for the parameter range which is studied above.

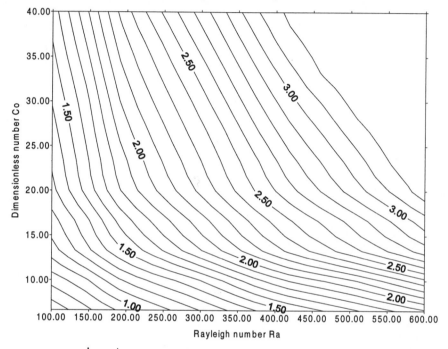

Fig. 13.10: $|\Psi_{min}|$ contours in the Ra-Co plane - postprocessing using results from model runs with FAST-C(2D)

The minimum value which the streamfunction takes on the grid cannot be estimated as easily. The dependence of $|\Psi_{min}|$ on Ra and Co is visualized by contour lines in Fig. 13.10.

Grid Péclet criteria were fulfilled in most model runs of the sensitivity analysis. For high Rayleigh numbers and low values of Co the criteria were not fulfilled. For Ra=600 the criteria were fulfilled only for Co=40. For Co=30 the criteria were not fulfilled in two blocks only; for Co=3.333 the Pe-values were above the margin in 2579 blocks. For Ra=100 the criterion was fulfilled for all values of Co.

Further variations will not be reported here. Readers may vary parameters themselves starting from the CD-ROM input-data file.

14 Desert Sedimentary Basins

14.1 System Description

Under certain conditions density driven flow patterns can be relevant in desert sedimentary basins:

- The basin is closed, i.e. there is no surface and no subsurface outlet for water. The only outlet is the atmosphere.

- Sufficient rainfall in mountainous regions surrounding the basin.

- Dry (arid) climatic conditions at the playa with high potential evaporation

Groundwater, recharged by precipitation in the mountains, flows toward the valley reaching regions of lower elevation with arid climate. If hydraulic conditions are artesian, water leaves the subsurface. Above the surface evaporation leads to heavy losses of fluid. What remains, are lakes with increased salt concentration. In case of high evaporation salt marshes may form or even salt crusts on the ground surface. Under arid climatic conditions near surface groundwater is lost to the atmosphere. Thus an artesian hydraulic regime is not a prerequisite for the formation of salt crusts.

The conditions listed above can be found in desert valleys all over the world. Fan e.a. (1997) list cases in the US in Utah, Nevada California and New Mexico and mention a chain of playas in the Australian desert. A further example is the Dead Sea Valley.

In the described situations density gradients caused by salinity changes are much higher than density gradients induced by temperature change. The following calculation was made by Tóth (1984) and recalled by Fan e.a. (1997). Under the assumption that the salinity changes from saturation to fresh water over the depth of 1500m the density gradient becomes 126 kg/m^3/km. The value significantly exceeds that for induced by the relatively high geothermal gradient of 4.9°C per 100m which is 18.4 kg/m^3/km.

It should be noted that in the described setting both gradients favor the onset of convection. They are not acting in different directions, as in the classical situation of double diffusive systems (see Brandt e.a. 1995) - in which salinity and temperature increase with depth. Convection can be observed for concentrations below saturation salinity. In Mono Lake Basin, California, Rogers/Dreiss (1995a) find evidence for convective motions in groundwater. The concentrations in the vicinity of the lake are about 18 ppt.

In the above mentioned cases the natural conditions lead to high salt concentrations in surface waters, in fluid and soil in the unsaturated zone and in the groundwater. Severe interaction of men may lead to similar conditions at much shorter time scales. The Aral Sea basin is an example where this is demonstrated at a relatively large scale. Huge output from streams for irrigation purposes are taken since the 50s. For the lack of surface inflow the Aral Sea shrunk to half of its area and a third of its volume. The salinity increased from 0.9 ppt in the 60s up to 35 ppt at present and is predicted to reach 43 ppt by the year 2015 (Bortnik 1992). In the groundwater the recharge of saline water from the Aral Sea will lead to convective motions.

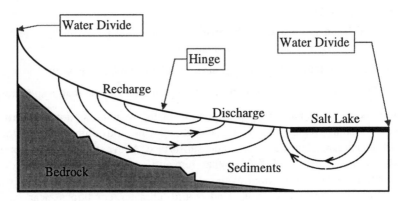

Fig. 14.1: Schematic view of conditions in sedimentary basins with hydrologic closure

In the following an idealized simple situation is discussed which is characterized by:

● Steady state conditions.

● Along a streamline in a horizontal plane flow can be represented by a 2D cross-section.

Fig. 14.2 gives an illustration of flow and transport boundary conditions in the streamfunction formulation. At three sides the system is closed. Below the boundary is the separating surface from the sediment layer to the underlying bedrock. At vertical edges the boundaries are laid along water divides. In real systems the sediment basin may not be separated from impermeable rock by a horizontal surface and water divides may not be vertical. For the hypothetical constellation discussed here it is assumed because of simplicity.

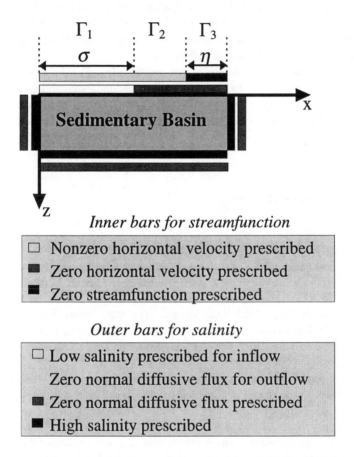

Fig. 14.2: Illustration of boundary conditions in streamfunction formulation

As in former chapters the no-flow situation is given by a zero Dirichlet boundary condition for the streamfunction and a Neumann condition for the transport variable. The first condition guarantees that there is no advective flux of salt across the boundary; the second guarantees no dispersive/diffusive flux.

The top edge of the model area needs to be divided in three parts: Γ_1, Γ_2 and Γ_3. There are two types of boundary conditions both for streamfunction and salinity. The division into two parts is not the same for flow and transport.

On the right side of the model a shallow salt lake is situated. The condition for normalized salinity is: $\theta=1$. Below the lake saltwater infiltrates in vertical direction. The flow condition is that the horizontal velocity components vanish: $\partial\Psi/\partial x=0$.

On the left side (see Fig. 14.2) there is a specified non-zero horizontal component of velocity. In terms of streamfunction that means: $\partial \Psi / \partial z = v_{bound}$. A Dirichlet condition for Ψ could be used alternatively if inflow, i.e. groundwater recharge, is known and used for the formulation. For transport a mixed condition can be specified: no dispersive flux in case of outflow boundaries and specified freshwater salinity ($\theta = 0$) in case of inflow.

In the intermediate part of the upper model edge, Γ_2, the mixed transport condition is connected with a zero horizontal flux boundary condition. This region is the main outflow region in the cross-section, but it is not necessarily identical with the outflow part. Note that recharge or discharge are calculated in the simulation run.

14.2 Numerical Modeling

Numerical modeling for the study of flow in desert sedimentary basins has been applied in few cases only. Rogers/Dreiss (1995a,b) take the SUTRA code written by Voss (1984) to simulate flow and salt transport in the Mono Lake basin, California. SUTRA again is applied by Fan e.a. (1997) in a study on Pilot Valley on the border between Utah and Nevada. The later simulation is in fact an extension of a former model by Duffy/Al-Hassan (1988). The models conditions are similar and only the differences will be mentioned here.

Rogers/Dreiss (1995a,b) use cylindrical coordinates, i.e. they consider the model cross-section as a part of a concentric system. Fan e.a. (1997) and Duffy/Al-Hassan formulate the model in cartesian coordinates. The boundary conditions at the three closed model edges are the same: Neumann type for pressure and concentration.

There are fundamental differences at the top model edge. Rogers/Dreiss (1995a,b) distinguish only two parts with different conditions: one could say they take Γ_2 with zero length (compare Fig. 14.2).

Fan e.a. (1997) and Duffy/Al-Hassan (1995) prescribe an inflow velocity which is derived from groundwater recharge. This is a Neumann condition for pressure and is equivalent to a Dirichlet condition for streamfunction. Rogers/Dreiss (1995a,b) instead chose a Dirichlet condition for pressure which can be transformed in a Neumann condition for streamfunction.

Both types of conditions seem to be reasonable. As both can be formulated for Ψ as well, the difference can be studied using a FAST-C(2D) model. For this chapter a model has been set up using FAST-C(2D) to simulate transient and steady state flow fields under the idealized conditions as noted in the preceding chapter.

Input parameters for a hypothetic example have been selected in accordance with Fan e.a. (1997). The relevant parameters and their values are shown here again in Table 14.1. An effective diffusivity is derived using the inflow velocity and dispersivity $\alpha := \alpha_L = \alpha_T$:

$$D = \alpha \frac{q}{\varphi} \qquad (14.1)$$

Of course the constant diffusivity assumption makes a difference to the velocity dependent situation assumed by Fan e.a. (1997).

Symbol	Parameter	Value	Phys. units
k	permeability	10^{-13}	m^2
$\Delta\rho$	density difference	150	kg/m^3
μ	viscosity	10^{-3}	$kg/m/s$
φ	porosity	0.35	1
D	diffusivity	$4.8 \cdot 10^{-7}$	m^2/s
q_{bound}	recharge velocity	$7.6 \cdot 10^{-2}$	m/a
H	height	1300	m
L	length	13000	m

Table 14.1: Relevant input parameters for modeling desert sedimentary basin

Using the FAST-C(2D) code the option for dimensionless parameters has been chosen[1]. The input values need to be transformed in dimensionless form - the outcome discretization is listed in Table 14.2.

Symbol	Parameter	Value (dimensionless)
H	height	1.
L	length	10.
σ	see Fig. 14.2	7.4
η	see Fig. 14.2	0.6
Ra	Rayleigh number	400.
v_{bound}	Boundary velocity	6.5

Table 14.2: Input data for the numerical model using FAST-C(2D)

---

[1] The input data for the model can be found on CD-ROM in the file *SedBasin.WTX*

Fig. 14.3 shows the outcome of the steady state numerical model. Salinity is represented by different gray patterns, ranging from 5% up to the 95% level. Flow is visualized by streamlines.

Beneath the top salt layer water sinks to the bottom. The saline fluid mixes with fresh water along horizontal passages. The brackish water with reduced density rises again and finally reaches the outflow region at the top of the model. The dispersion zone is rather small.

The described situation is the result from a steady state model. The entire aquifer below the salt lake or salt crust is filled with highly saline water. The mixing takes place in horizontal direction mainly.

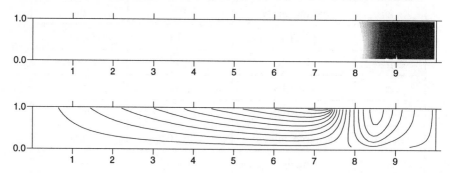

Fig. 14.3: Numerical output from FAST-C(2D) steady state model; top: salinity distribution; bottom: streamlines

The situation is different in a transient situation, when fresh waters are overlain by saline waters. It can be expected that convective motions are stronger in unsteady situations and that mixing takes place mainly during the vertical passage of fluid towards the aquifer bottom.

Concluding Remark

This is the end of the book but not the end of problems with density-driven flow.

The presented numerical approach can be applied to several types of problems but has limitations. As author and programmer of the codes I am tempted to work to overcome the constraints. The program would become more complex and the number of parameters would increase. I would have to document the new implementations and describe them in the help-system. Perhaps I would end up with another volume.

The applications described are mostly hypothetical cases. The step to real field applications is big. Inhomogeneities have to be considered and anisotropies and irregular boundaries and local grid-refinements and ...! This could not be done in what is a textbook after all.

There are applications which are not mentioned. When I started the book I wanted to include the 'island problem', where the size and location of the fresh water lens below an island is studied. This is surely not the only problem which got lost.

The problems are not solved but I hope that, using the book and the software, problems can be formulated at a higher level. This is my hope.

The other hope is that I will be able to maintain and extend the software in the future. If you find errors in book or software please give a note to my e-mail address:

holzbecher@igb-berlin.de

Bugs and errors will be reported in the internet. You will be informed about all new developments and extensions concerning programs, descriptions and applications on the Web-site:

http://www.igb-berlin.de/www/abt1/holzbecher/holzbecher_e.htm

References

Abraham G./ van Os A.G./Verboom G.K., Mathematical modeling of flows and transport of conservative substances: requirements for predictive ability, in: Fischer (ed.), Transport models for inland and coastal waters, Acad. Press, New York, 1-31, 1981

Ackerer P./ Mose R./ Oltean C./ Siegel P., Experimental and numerical study of density flow, oral presentation at the mini-symposium on 'Modelling density-driven flow in porous media', Heidelberg, 1995

Aidun C.K., Stability of convection rolls in porous media, in: Bifurcation Phenomena in Thermal Processes and Convection, ASME, HTD-Vol. 94 and AMD-Vol. 89, 31-36, 1987

Aidun C.K./ Steen P.H., Transition to oscillatory convection in a fluid-saturated porous medium, *J. Thermophys.*, Vol. 1, No.3, 268-273, 1987

Amari B./ Vasseur P./ Bilgen E., Natural convection of non-Newtonian fluids in a horizontal porous layer, *Wärme- und Stoffübertragung*, Vol. 29, 185-193, 1994

Arlt H., A hydrogeological study of the Nile delta aquifer with emphasis on saltwater intrusion in the Northern Area, Techn. Univ. Berlin, Inst. für Wasserbau und Wasserwirtschaft, Mitteilung 130, 185p, 1995

Aszódi A., Experimentelle und numerische Untersuchungen der Erwärmung von Kies-Wasser-Mischungen in beheizten Behältern, Verein für kernverfahrenstechnik und Analytik Rossendorf e.V., VKTA - 07, 61p, 1996

Atkinson R./ Herbert A.W./ Jackson C.P./ Robinson P.C., NAMMU user guide, Dep. of Environment, Report DOE/RW/85.065, 50p, 1985

Axelsson, O./ Gustaffson, I., A modified upwind scheme for convective transport equations and the use od a conjugate gradient method for the solution of nonsymmetric systems of equations, *J. Inst. Math. Applic.*, Vol. 23, 321-337, 1979

Axelsson O./ Vassilevski P.S., A survey of multilevel preconditioned iterative methods, *BIT*, Vol. 29, 769-793, 1989

Aziz K./ Settari A., Petroleum reservoir simulation, Applied Science Publishers, London, 476p, 1979

Badon Ghyben W., Nota in verband met de voorgenomen putboring nabij Amsterdam, *Tijdschrift van het Koninklijk Institut voor Ingenieurs*, The Hague, 8-22, 1889

Bastian P./ Birken K./ Johannsen K./ Lang S./ Neuß N./ Rentz-Reichert H./ Wieners C., UG - a fexible software toolbox for solving partial differential equations, *Computing and Visualization in Science*, Vol. 1, No. 1, 27-40, 1997

Bau H.H./ Torrance K.E., Low Rayleigh number thermal convection in a vertical cylinder filled with porous materials and heated from below, *ASME J. Heat Transfer*, Vol. 104, 166-172, 1982

Baumann R., Untersuchungen zum anthropogenen Einfluß auf die Salzwasserinvasion im Nildelta, Techn. Univ. Berlin, Inst. für Wasserbau und Wasserwirtschaft, Mitteilung 129, 113p, 1995

Baumann R./ Moser H., Modellierung der Meerwasserinvasion im Delta arider und semiarider Gebiete am Beispiel des Nildeltas, *Z. dt. geol. Ges.*, Nr. 143, 316-324, 1992

Bear J., Hydraulics of Groundwater, Mc Graw Hill, New York, 569p, 1976

Bear J., Flow through porous media, Elsevier, New York, 764p, 1972

Bear J./ Cheng A. H.-D./ Herrera I./ Sorek S./ Quazar D., Seawater intrusion in coastal aquifers, Kluwer Publ., Dordrecht, to appear (1998)

Bear J./ Verruijt A., Modeling groundwater flow and pollution, D. Reidel Publ., Dordrecht, 414p, 1987

Beck J.L., Convection in a box of porous material saturated with fluid, *Phys. Fluids*, Vol. 15, 1377-1383, 1972

Bejan A., Convective heat transfer in porous media, in: S.Kakaç/ Shah R.K./ Aung W. (eds), Handbook of Single-Phase Convective Heat Transfer, Wiley, New York, 1987

Bejan A./ Tien C.L., Natural convection in a horizontal porous medium subjected to an end-to-end temperature difference, ASME *J. Heat Transfer*, Vol. 100, 191-198, 1978

Bénard H., Les tourbillons cellulaires dans nappe liqiude transportant de la chaleur par convections en regime permanent, *Rev. Gen. Sci. Pures Appl. Bull. Assoc.* 11, 1261-1271, 1309-1328, 1900

Bennett G.D./ Mundorff M.J./ Hussain S.A., Electric-analog studies of brine coning beneath fresh-water wells in the Punjab region, U.S. Geol. Survey, Water Supply Paper 1608-J, J1-J31, 1968

Bergmann L./ Schaefer C., Mechanik Akustik Wärme - Lehrbuch der Experimentalphysik, Band I, Walter de Gruyter, Berlin, 902p, 1990

Bhattacharjee J.K., Convection and Chaos in Fluids, World Scientific, Singapore, 243p, 1987

Bjørlykke K./ Mo A./ Palm E., Modelling of thermal convection in sedimentary basins and its relevance to diagenetic reactions, *Marine and Petroleum Geology*, Vol. 5, 338-351, 1988

Blake K.R./ Bejan A./ Poulikakos D., Natural convection near 4°C in a water saturated porous layer heated from below, *Int. J. Heat Mass Transfer*, Vol. 27, No.12, 2355-2364, 1984

Blakeley M.R./ O'Sullivan M.J., Modeling of production and recharge in Wairakei, *Proc. of the Pacific Geothermal Conf.*, 23-31, 1982

Bodvarsson G.S./ vonder Haar S./ Wilt M./ Tsang C.F., Preliminary studies of the reservoir capacity and the generating potential of the Baca geothermal field, *Water Res. Res.*, Vol. 18, No.6, 1713-1723, 1982

Bories S., Natural convection in porous media, in: Bear J./ Corapcioglu M.Y. (eds), Advances in Transport Phenomena in Porous Media, NATO ASI Series E: Applied Sciences, No.128, Martinus Nijhoff Publ., Dordrecht, 77-142, 1987

Bories S./ Thirriot C., Échanges thermiques et tourbillions dans une couche poreuse horizontale, *La Houille Blanche*, No.3, 237-245, 1969

Bortnik V.N./ Kuksa V.I./ Tsytsarin A.G., Present status and possible future of the Aral Sea, *Post-Soviet Geography*, Vol. 33, No.5, 315-322, 1992

Bouhlila R., Modelisation couplee d'ecoulement et de transport en milieux poreux: cas des fortes densités, Atelier International 'Application des modeles mathematiques a l'evaluation des modifications de la qualite de l'eau', Proceedings, 166-179, 1990

Brandt A./ Fernando H.J.S., Double-diffusive convection, Am. Geophys. Union, Geophys. Monograph 94, 334p, 1995

Brown D.A./ Lemons J., Scientific certainty and the laws that govern location of a potential high-level nuclear waste repository, *Environmental Management*, Vol. 15, No.3, 311-319, 1991

Buretta R.J./ Berman A.S., Convective heat transfer in a liquid saturated porous layer, *ASME J. of Appl. Mech.*, Vol. 43, 249-253, 1976

Bütow E., Salt water distribution in a saturated porous medium, Proposal for a test problem HYDROCOIN level 1 case 5, HYDROCOIN project, 6p, 1984

Bütow E. / Holzbecher E., Proposal for definition of HYDROCOIN level 3 case 4: sensitivity analysis for the flow over a salt dome, HYDROCOIN project, 5p, 1986

Bütow E./ Brühl G./ Gülker M./ Heredia L./ Lütkemeier-Hosseinipour S./ Naff R./ Struck S., Modellrechnungen zur Ausbreitung von Radionukliden im Deckgebirge, Abschlußbericht 'Projekt Sicherheitsstudien Entsorgung', Fachband 18, Berlin, 1985

Bütow E./ Holzbecher E., On the modeling of groundwater flow under the influence of salinity, in: Jousma G./ Bear J./ Haimes Y.Y./ Walter F. (eds), IGWMC-Conference on Groundwater Contamination (Use of models in decision- making), Proceedings, Amsterdam, 263-272, 1987

Caltagirone J.-P., Thermoconvective instabilities in a horizontal porous layer, *J.Fluid Mech.*, Vol. 72, No.2, 269-287, 1975

Caltagirone J.-P., Convection in a porous medium, in: Zierep/ Oertel (eds), Convective transport and instability phenomena, G.Braun, Karlsruhe, 199-232, 1982

Caltagirone J.-P./ Fabrie P., Natural convection in a porous medium at high Rayleigh numbers, *Eur. J. Mech. B/Fluids*, No.3, 207-227, 1989

CFEST, see: Gupta e.a. 1982

Canuto C./ Hussaini M.Y./ Quarteroni A./ Zang T.A., Spectral methods in fluid dynamics, Springer Publ., New York, 557p, 1987

Chandrasekhar S., Hydrodynamic and hydromagnetic stability, New York, 654p, 1961

Cheng P.: Heat transfer in geothermal systems, *Advances in Heat Transfer*, Vol. 14, 1-105, 1978

Cheng P./Lau K.H., The effect of steady withdrawal of fluid in confined geothermal reservoirs, 2. *UN-Symp. On Development and Use of Geothermal Reservoirs*, Section VII, 1591-1598, 1975

Close D.J./ Symons J.G./ White R.F., Convective heat transfer in shallow gas-filled porous media: experimental investigation, *Int. J. Heat Mass Transfer*, Vol. 28, No.12, 2371-2378, 1985

Cole C.R./ Nicolson T.J./ Davis P./ McCartin T.J., Lessons learned from the HYDROCOIN experience, in: SKI (Swedish Kärnkraftinspektion), *GEOVAL-87 symposion*, Proceedings Vol. 1, Stockholm, 269-285, 1988

Combarnou M.A./ Bories S.A., Hydrothermal convection in saturated porous media, *Adv. Hydrosci.*, Vol. 10, 231-307, 1975

Cooper H.H., A hypothesis concerning the dynamic balance of fresh water and salt water in a coastal aquifer, *J. Geophys. Res.,* Vol. 64, No.4, 461-467, 1959

Cordier E./ Goblet P., Contribution to the analysis of the RIVM brine migration experiments, INTRAVAL level 1, Commissariat à l'Energie Atomique, Report LHM/RD/91/8, 12p, 1991

Courant L./ Friedrichs K.O./ Lewy H., Über die partiellen Differentialgleichungen der mathematischen Physik, *Math. Ann.* 100, 1928

Cussler E.L., Diffusion: mass transfer in fluid systems, Cambridge Univ. Press, New York, 525p, 1984

Dagan G./ Bear J., Solving the problem of local interface upconing in a coastal aquifer by the method of small perturbations, *J. of Hydr. Res.*, Vol. 6, No.1., 1968

Dei H., Studies on dispersion of salts in coastal aquifers (1) - dispersion of salts caused by oscillating interface between saltwater and freshwater in porous media, *Jap. J. Limnol.,* Vol. 39, No.3, 112-117, 1978

Diersch H.-J.G., Primitive Variable Finite Element Solutions of Free Convection Flows in Porous Media, *ZAMM 61*, 325-337, 1981

Diersch H.-J.G., FEFLOW - Physikalische Modellgrundlagen, WASY Gmbh, Berlin, 78p, 1996

Diersch H.-J.G./ Nillert P., Saltwater intrusion processes in groundwater: novel computer simulations, field studies and interception techniques, Int. Symp. on Groundwater Monitoring and Management, Dresden, Complex IV, Paper 13, 17p, 1987

Diersch H.-J.G./ Prochnow D./ Thiele M., Finite-element analysis of dispersion-affected saltwater upconing below a pumping well, *Appl. Math. Modelling*, Vol. 8, 305-312, 1984

Donaldson I.G., The simulation of geothermal systems with a simple convective model, UN Symp. On Development and Utilisation of Geothermal Resources, I/10-VII/21, 1970

Dorsey N.E. Properties of ordinary water substance, New York, 1940

Duffy C.J./ Al-Hassan S., Groundwater circulation in a closed desert basin: topographic scaling and climatic forcing, *Water Res. Res.*, Vol. 24, No. 10, 1675-1688, 1988

DWDM - A model for calculating effects of liquid waste disposal in deep saline aquifers, US-Geol. Survey, Water Res. Investig. 76-61, 1976

van Dyke M., An Album of Fluid Motion, The Parabolic Press, Stanford, 176p,1982

Eisenberg N.A./ Alexander D.H./ Lee W.W.-L., The role of validation activities in the U.S. Dep. of Energy's office of civilian radioactive waste management program (Draft), in: SKI (Swedish Kärnkraftinspektion), *GEOVAL-87 Symposion*, Proceedings Vol. 1, Stockholm, 27-40, 1988

Elder J.W., Numerical experiment with free convection in a vertical slot, *J.Fluid Mech.*, Vol. 24, 823-843, 1966

Elder J.W., Steady free convection in a porous medium heated from below, *J. Fluid Mech.*, Vol. 27, 29-48, 1967a

Elder J.W., Transient convection in a porous medium, *J. Fluid Mech.*, Vol. 27, No.3, 609-623, 1967b

Elder J.W., The unstable thermal interface, *J. Fluid Mech.*, Vol. 32, No.1, 69-96, 1968

Ene H.I./ Poliševski D., Thermal flow in porous media, Reidel Publ. Comp., Dordrecht, 194p, 1987

Epherre J.F., Criterion for the appearance of natural convection in an anisotropic porous layer, *Int. Chem. Engng.*, Vol. 17, 615-616, 1975

Essaid H.I., A comparison of the coupled fresh water-salt water flow and the Ghyben-Herzberg sharp interface approaches to modeling of transient behavior in coastal aquifer systems, *J. of Hydrology*, Vol. 86, 169-193, 1986

Evans D.G./ Nunn J.A., Free thermohaline convection in sediments surrounding a salt column, *J. Geophys. Res.*, Vol. 94, B9, 12413-12422, 1989

Evans D.G./ Nunn J.A./ Hanor J.S., Mechanisms driving groundwater flow near saltdomes, *Geophys. Res. Lett.*, Vol. 18, No.5, 927-930, 1991

Evans D.G./ Raffensperger J.P., On the stream function for variable-density groundwater flow, *Water Res. Res.*, Vol. 28, No.8, 2141-2145, 1992

Evans D.J., The use of pre-conditioning in iterative methods for solving linear equations with symmetric positive definite matrices, *J. Inst. Maths Applics*, Vol. 4, 295-314, 1967

Fan Y./ Duffy C.J./ Oliver D.S., Density-driven groundwater flow in closed desert basins: field investigations and numerical experiments, *J. of Hydrology*, Vol. 196, 139-184, 1997

FEFLOW - see: Diersch 1996

Fife P.C., The Bénard problem for general fluid dynamical equations and remarks on the Boussinesq approximation, *Indiana Univ. Math. J.*, Vol. 20, No.4, 303-326, 1970

Fletcher R., Conjugate gradient methods for indefinite systems, in: Lecture Notes in Mathematics, No. 506, Springer Publ., Berlin, 73-89, 1976

Freeze R.A./Cherry J.A., Groundwater, Prentice-Hall, Englewood Cliffs, 604p, 1979

Frind E.O., Seawater intrusion in continuous coastal aquifer-aquitard systems, in: Wang S.Y. / Brebbia C.A./ Alonso C.V./ Gray W.G./ Pinder G.F. (eds), Fin. Elem. in Water Res., 3rd Intern. Conf., 2.177-2.198, 1980

Frind E.O., Simulation of long-term transient density-dependent transport in groundwater, *Adv. Water Res.*, Vol. 5, 73-88, 1982

Frind E.O./ Matanga G.B., The dual formulation of flow for contaminant transport modeling 1: review of theory and accuracy aspects, *Water Res. Res.*, Vol. 21, No. 2, 159-169, 1985

Galeati G./ Gambolati G./ Neuman S.P., Coupled and partially coupled Eulerian-Lagrangian model of freshwater-seawater mixing, *Water Res. Res.*, Vol. 28, No.1, 149-165, 1992

Garg S.K./ Kassoy D.R., Convective heat and mass transfer in hydrothermal systems, in: Rybach L./ Muffler L.J.P. (eds): *Geothermal systems: principles and case histories*, Chichester, 37-76, 1981

Garven G., A hydrogeologic model for the formation of the giant oil sand deposits of the western Canada sedimentary basin, *Am. J. Sci*, Vol.289, 105-166,1989

van Genuchten M. Th., Analytical solutions for chemical transport with simultaneous adsorption, zero-order production and first-order decay, *J. of Hydrology*, Vol. 49, 213-233, 1981

van Genuchten M. Th./ Wierenga P.J., Mass transfer studies in sorbing porous media I. Analytical studies, *Journal of the Soil Science Soc. of America*, Vol. 40, No.4, 473-480, 1976

Grohman A., Bedeutung und Bestimmung der elektrischen Leitfähigkeit, in: Aurand (ed.), Die Trinkwasserverordnung, Erich Schmidt Publ., Berlin, 308-309, 1987

Gupta S.K./ Kincaid C.T./ Meyer P.R./ Newbill C.A./ Cole C.R., CFEST-multi-dimensional finite-element code for the analysis of coupled fluid, energy and solute transport, Pacific Northwest Laboratory-4260, Richland, 1982

Guvanasen V., Comments on case 5, level 1 - HYDROCOIN, communication to HYDROCOIN secretariat, 2p, 1985

Häfner F./ Sames D./ Voigt H.-D., Wärme- und Stofftransport, Springer Publ., Heidelberg, 626p, 1992

Hassanizadeh S.M., Verification and validation of coupled flow and transport models, in: SKI (Swedish Kärnkraftinspektion), *GEOVAL-90 symposion*, Proceedings, Stockholm, 1990a

Hassanizadeh S.M., Experimental study of coupled flow and mass transport: a model validation exercise, in: K. Kovar (ed.), *Intern. Conf. on Calibration and Reliability in Groundwater Modelling (ModelCARE)*, Den Haag, Proceedings, IAHS-Publication No.195, 241-250, 1990b

Hassanizadeh S.M./ Gray W.G., General conservation equations for multi-phase systems: 1. averaging procedures, *Adv. in Water Res.*, Vol. 2, 131-144, 1979a

Hassanizadeh S.M./ Gray W.G., General conservation equations for multi-phase systems: 2. mass, energy and entropy equations, *Adv. in Water Res.*, Vol. 2, 191-203, 1979b

Hassanizadeh S.M./ Gray W.G., General conservation equations for multi-phase systems: 3. constitutive theory for porous media, *Adv. in Water Res.*, Vol. 3, 25-40, 1980

Hele-Shaw H.S.J., *Trans. Inst. Nav. Arch.*, Vol. 40, 21ff, 1898

Helmig R., Multiphase flow and transport processes in the subsurface, Springer Publ., Berlin, 367p, 1997

Hem J.D., Study and interpretation of the chemical characteristics of natural waters, U.S. Geol. Survey, Water-Supply Paper 1473, Washington D.C., 363p, 1970

Henry H.R., Salt intrusion into coastal aquifers, Intern. Ass. of. Scient. Hydrology Publ., No. 52, 478-487, 1960

Henry H.R., Interfaces between salt water and fresh water in coastal aquifers, Geol. Survey Water-supply Paper 1613-C, C35-C70, 1964

Herbert A.W./ Jackson C.P./ Lever D.A., Coupled groundwater flow and solute transport with fluid density strongly dependent upon concentration, Water Res. Res., Vol. 24, 1781-1795, 1988

Heredia L./ Holzbecher E., Model for variable-density-flow, Proposal for test problem HYDROCOIN level 2 case 2, 70p, 1986

Herzberg A., Die Wasserversorgung einiger Nordseebäder, Journal für Gasbeleuchtung und Wasserversorgung, Vol. 44, 815-819, 842-844, 1901

Hestenes M.R., Conjugate direction methods in optimization, Springer Publ., New York, 325p, 1980

Hill M.C., A comparison of coupled freshwater-saltwater sharp-interface and convective-dispersive models of saltwater intrusion in a layered aquifer system, Comp. Meth. in Water Res. VII. Intern. Conf., 211-216, 1988

Holst, P.H./ Aziz, K., Transient three-dimensional convection in confined porous media, *Int. J. Mass Heat Transfer*, Vol. 15, 73-89, 1972

Holzbecher E., Supplement to specification of HYDROCOIN level 2 case 2, internal HYDROCOIN paper, 1p, 1986

Holzbecher E., Numerische Modellierung von Dichtestroemungen im poroesen Medium, Techn. Univ. Berlin, Inst. für Wasserbau und Wasserwirtschaft, Mitteilung Nr.117, 160p, 1991a

Holzbecher E., Zur Modellierung von Dichteströmungen in porösen Medien, in: Holzbecher E./ Nützmann G. (eds) Modellierung von Strömung und Grundwasser im Grundwasser, Techn. Univ. Berlin, Inst. für Wasserbau und Wasserwirtschaft, Mitteilung Nr.120, 93-110, 1991b

Holzbecher E., Modeling the Yufuin geothermal system in Kyushu (Japan), unpublished, 1993

Holzbecher E., Modeling of saltwater upconing, in: Wang S. (ed.) II. Int. Conf. Hydro-Science and Hydro-Engin., Proceedings Vol. 2, Part A, Beijing, 858-865, 1995

Holzbecher E., Modellierung dynamischer Prozesse in der Hydrologie: Grundwasser und ungesättigte Zone, Springer Publ., Heidelberg, 1996

Holzbecher E., Numerical studies on thermal convection in cold groundwater, *Int. J. Heat Mass Transfer*, Vol. 40, No.3, 605-612, 1997a

Holzbecher E., Decay chain equilibria in a two-phase environment with linear sorption, *J. of Environmental Hydrology*, Vol. 5, 1997b

Holzbecher E., Remarks on the paper by E.J. Rykiel, entitled: 'testing ecological models: the meaning of validation', *Ecol. Model.*, Vol. 102, 375-376, 1997c

Holzbecher E./ Baumann R., Numerical simulations of saltwater intrusion into the Nile Delta Aquifer, in: Peters A./ Wittum G./ Herrling B./ Meissner U./ Brebbia C.A/ Gray W.G./ Pinder G.F. (eds), Comp. Meth. in Water Res. X, Proc. Vol. 2, Kluwer Publ., Dordrecht, 1011-1018, 1994

Holzbecher E./ Heinl M., Anisotropy and dispersivity effects on saltwater upconing, Comp. Methods and Water Ressources, Beirut, 117-126, 1995

Holzbecher E./ Yusa Y., Numerical experiments on free and forced convection in porous media, *Int. J. Heat Mass Transfer*, Vol. 38, No. 11, 2109-2115, 1995

Horne R.N./ O'Sullivan M.J., Oscillatory convection in a porous medium heated from below, *J.Fluid Mech.*, Vol. 66, 339-352, 1974

Hosokawa, T., Jinno, K. and Momii, K., Estimation of transverse dispersivity in the mixing zone of fresh-salt groundwater, in: Kovar, K. (ed.) Calibration and Reliability in Groundwater Modelling, IAHS Publ. No. 195, 149-158, 1990

Hossain S./ Weber P., Prediction of thermally induced flow in the sealing of boreholes containing high level nuclear wastes, in: Hossain S./ Podtschaske T./ Rimkus D. Stelte N,/ Storck R./ Weber P. (eds), Einzeluntersuchungen zur Radionuklidfreisetzung aus einem Modellsalzstock, Projekt Sicherheitsstudien Entsorgung PSE, Fachband 15, 195-260, 1985

Hubbert M.K., The theory of ground-water motion, *J. of Geology*, Vol. XLVIII, No.8, 785-938, 1940

Huisman I.L., La formation des cones d'eau saumatre, IUGG, IAHS-Publ. No.37, 146-150, 1954

Huyakorn P.S./ Pinder G.F., Computational Methods in Subsurface Flow, Academic Press, New York, 473p, 1983

Huyacorn P.S./ Anderson P.F./ Mercer J.W./ White H.O., Saltwater intrusion in aquifers: development and testing of a three-dimensional finite element model, *Water Res. Res.*, Vol. 23, No.2, 293-312, 1987

HYDROCOIN - The International HYDROCOIN Project, Level 1: code verification, OECD, Paris, 198p, 1988

HYDROCOIN - The International HYDROCOIN Project, Level 2: model validation, OECD, Paris, 194p, 1990

IAEA, Radioactive waste management glossary, TECDOC-447, Vienna, 1988

Ingebritsen S.E./ Sorey M.L., A quantitative analysis of the Lassen hydrothermal system, *Water Res. Res.*, Vol. 21, No.6, 853-868, 1985

Inouchi K./ Kakinuma T., Behaviour of coastal groundwater in layered aquifers interbedded with semi-permeable layer, *Jap. J. Limnology*, Vol. 53, No.3, 187-196, 1992

INTERA Environmental Consultants, Revision of the documentation for a model for calculating effects of liquid waste disposal in deep saline aquifers, U.S. Geol. Survey, Water Res. Publ. 79-96, 72p, 1979

INTRAVAL - Phase1, Case 13, Final report, 54p, 1991

INTRAVAL - The International INTRAVAL Project, Phase 1, Summary Report, OECD, Paris, 135p, 1994

Jackson M.P.A./ Talbot C.J., External shapes, starin rates and dynamics of salt structures, *Geol. Soc. Of Am. Bull.*, Vol. 97, No.3, 305-323, 1986

Jennings A.A., The impact of phase exchange transfers and unsteady hydraulic conditions on saltwater intrusion fronts, in: Laible J.P./ Brebbia C.A./ Gray W./ Pinder G. (eds), Finite Elements in Water Resources V, Springer Publ., Berlin, 143-152, 1984

Johns R.T./ Rivera A., Comment on 'Dispersive transport dynamics in a strongly coupled groundwater-brine flow system' by C.M.Oldenburg and K. Pruess, *Water Res. Res.*, Vol. 32, No.11, 3405-3410, 1996

Jonsson T./ Cattan I., Prandtl number dependence of natural convection in porous media, *ASME J. of Heat Transfer*, Vol. 109, 371-377, 1987

Joseph D.D., Stability of fluid motions II, Springer Publ., Berlin, 274p, 1976

de Josselin de Jong G., Singularity distributions for the analysis of multiple-fluid flow through porous media, *J. Geophys. Res.*, Vol. 65, No.11, 3739-3758, 1960

Jovkov B., Salt and fresh water interaction in aquifers and its control, Int. Symp. On Groundwater Monitoring and Management, Dresden, Complex III, Paper 11, 15p, 1987

JSME (Jap. Soc. of Mech. Eng.), Steam tables, Tokyo, 1968

Kaneko T./ Mohtadi M.F./ Aziz K., An experimental study of natural convection in inclined porous media, *Int. J. Heat Mass Transfer*, Vol. 17, 485-496, 1974

Kasch M., Theoretical investigation on gas flow through porous media due to large pressure gradients, *J. of Porous Media*, Vol. 1, No.3, 253-260, 1997

Kassoy D.R., Heat and Mass Transfer in Models of Undeveloped Geothermal Fields, *Proc. 2. UN-Symp. on Development and Use of Geothermal Resources*, Section VII, San Francisco, 1707-1711, 1975

Kassoy D.R./ Zebib A., Variable viscosity effects on the onset of convection in porous medium, *Physics of Fluids*, Vol. 18, 1649-1651, 1976

Kastenberg W.E./ Gratton L.J., Hazards of managing and disposing of nuclear waste, *Physics Today*, Vol. 50, No.6, 41-47, 1997

Katto Y./ Masuoka, Criterion for the onset of convective flow in a fluid in a porous medium, *Int. J. Heat Mass Transfer*, Vol. 10, 297-309, 1967

Kimura S./ Schubert G./ Straus J.M., Route to chaos in porous medium thermal convection, *J. Fluid Mech.*, Vol. 166, 305-324, 1986

Kinzelbach W., Groundwater modelling, Elsevier, Amsterdam, 333p, 1986

Kinzelbach W., Numerische Methoden zur Modellierung des Transports von Schadstoffen im Grundwasser, Oldenbourg, München, 317p, 1987

Kitaoka K./ Kikkawa K., Thermal groundwater in deep stratified reservoirs in Oita prefecture, unpublished, 1992

Knudsen W.C., Equations on fluid flow through porous media - incompressibel fluid of varying density, *J. Geophys. Res.*, Vol. 67, No.2, 733-737, 1962

Kohout F.A./ Klein H., Effect of pulse recharge on the zone of diffusion in the Biscayne aquifer, in: Artificial Recharge and Management of Aquifers, Symposion of Haifa, IAHS Publ. No.72, 252-270, 1967

Kolditz O., Benchmarks for numerical groundwater simulations, in: Diersch H.J., FEFLOW User's Manual, Release 4.20, WASY, Berlin, 5.1-5.129, 1994

Kolditz O./ Ratke R./ Diersch H.-J./ Zielke W., Coupled groundwater flow and transport: 1. Verification of variable density flow and transport models, *Adv. in Water Res.*, Vol. 21, No.1, 27-46, 1998

Konikow L.F./ Bredehoeft J.D., Computer model of two-dimensional solute transport and dispersion in ground water, U.S. Geol. Survey, Techniques of Water-Res. Invest., Book 7, Chapter C2, 37p, 1978

Konikow L.F./ Sanford W.E./ Campbell P.J., Constant-concentration boundary condition: lessons from the HYDROCOIN variable-density groundwater benchmark problem, *Water Res. Res.*, Vol. 33, No.10, 2253-2261, 1997

Krige L.J., Borehole temperatures in the Transvaal and Orange Free State, Royal Soc. London Proc., Ser. A, Vol. 173, 450-474, 1939

Kukulka D.J./ Gebhart B./ Mollendorf J.C., Thermodynamic and transport properties of pure and saline waters, *Adv. Heat Transfer*, Vol. 18, 325-363, 1987

Lamb H., Hydrodynamics, Cambridge University Press, 738p, 1963

Lantz R.B., Quantitative evaluation of numerical diffusion (truncation error), *Soc. of Petr. Eng. J.*, 315-320, 1971

Lapwood E.R., Convection of a fluid in a porous medium, *Proc. Camb. Phil. Soc. A* 225, 508-521, 1948

Larsson A., The international projects INTRACOIN, HYDROCOIN and INTERVAL, *Adv. in Water Res.*, Vol. 15, 85-87, 1992

Lee C./ Cheng R.T., On seawater encroachment in coastal aquifers, *Water Res. Res.*, Vol. 10, No.5, 1039-1043, 1974

Leendertse J.J., Discussion-contribution in: Fischer (ed.), Transport models for inland and coastal waters, New York, 31-35, 1981

Leijnse A., Modelling the flow of groundwater in the vicinity of a salt dome, *Waste Management*, Vol. 3, 203-206, 1985a

Leijnse A., Optional ammendments to level 1 case5 with comments from the secretariate, HYDROCOIN newsletter, 2p, 1985b

Leijnse A., Comparison of solution methods for coupled flow and transport in porous media, in: in: Russell T.F./ Ewing R.E./ Brebbia C.A./ Gray W.G./ Pinder G.F. (eds), Comp. Meth. in Water Res. IX, Proceedings Vol. 2, Comp. Mech, Publ., Southampton, 273-280, 1992

Leijnse A., Some aspects of modelling density dependent flow and transport in porous media, Oral presentation at the mini-symposium on 'Modelling density-driven flow in porous media', Heidelberg, 1995

Leijnse A./ Hassanizadeh S.M., Verification of the METROPOL code for density dependent flow in porous media - HYDROCOIN project, level 2 case 2, Report 728528002, RIVM, Bilthoven, 1989

Leijnse A./ Oostrom M., The onset of instabilities in the numerical simulation of density-driven flow in porous media, in: Comp. Meth. in Water Res. X, Peters A./ Wittum G./ Herrling B./ Meissner U./ Brebbia C.A/ Gray W.G./ Pinder G.F. (eds) Proc. Vol. 2, Kluwer Publ., Dordrecht, 489-496, 1994

Leismann H.M./ Frind E.O., A symmetric-matrix time integration scheme for the efficient solution of advection-dispersion problems, *Water Res. Res.*, Vol. 25, No.6, 1133-1139, 1989

Lever D.A./ Jackson C.P., On the equations for the flow of concentrated salt solution through a porous medium, Dep. of Environm., Report DOE/RW/85.100m, 14p, 1985

Li Y.-H./ Gregory S., Diffusions in sea water and in deep sea sediments, *Geochim. et Cosmochim. Acta*, Vol. 38, 703-714, 1974

Lippmann M.J./ Bodvarsson G.S., Numerical studies of the heat and mass transport in the Cerro Prieto geothermal field, *Water Res. Res.*, Vol. 19, No.3, 753-767, 1983

Lorenz E.N., Deterministic Nonperiodic Flow, *J. Atmos. Sci.*, Vol. 20, 130-141, 1963

Masuoka T., Heat transfer by free convection in a porous layer heated from below, *Heat Transfer - Jap. Res.*, Vol. 1, No.1, 39-45, 1972

McCombie C., Nuclear waste management worldwide, *Physics Today*, Vol. 50, No.6, 56-62, 1997

Meis Th./ Marcowitz U., Numerische Behandlung partieller Differentialgleichungen, Springer Publ., Berlin, 452p, 1978

Mercer J.W./ Pinder G.F., Galerkin finite element simulation of a geothermal resrvoir, *Geothermics*, Vol. 2, No. 3&4, 81-88

METROPOL - see: Sauter

Mihaljan J.M., The rigorous exposition of the Boussinesq approximations applicable to a thin layer of fluid, *Astrophysical J.*, Vol. 136, 1126-1133, 1962

Molenaar J., Multigrid for a groundwater flow problem, Delft University, Report of the Faculty of Mathematics and Informatics, No.93-29, 1993

Momii, K., Hosokawa, T., Jinno, K. and Itoh, T., Estimation method of transverse dispersivity based on vertical salt concentration distribution in coastal aquifers, *Proc. of Japan Society of Civil Engineers* 411 (II-12), 45-53, 1989

Morris C.W./ Campbell D.A, Geothermal reservoir energy recovery - a three dimensional simulation study on the East Mesa Field, *J. of Petr. Techn.*, Vol. 33, 735-742, 1981

Morland L.W./ Zebib A./ Kassoy D.R., Variable property effects on the onset of convection in an elastic porous matrix, *Physics of Fluids,* Vol. 20, 1255-1259, 1977

Morrison H.L., Preliminary measurements relative to the onset of thermal convection currents in unconsolidated sands, *J. Appl. Phys.*, Vol. 18, 849-850, 1947

Morrison H.L./ Rogers F.T./ Horton C.W., Convection currents in porous media - observation of conditions at onset of convection, *J. Appl. Phys.*, Vol. 20, 1027-1029, 1949

Morrison H.L./ Rogers F.T., Significance of flow-patterns for initial convection in porous media, *J. Appl. Phys.*, Vol. 23, 1058-1059, 1952

Moser H., Einfluß der Salzkonzentration auf die hydrodynamische Dispersion im porösen Medium, Techn. Univ. Berlin, Inst. für Wasserbau und Wasserwirtschaft, Mitteilung 128, 95p, 1995

Motz L.H., Salt-water upconing in an aquifer overlain by a leaky confining bed, *Groundwater*, Vol. 30, No.2, 192-198, 1992

Mullin T., A multiple bifurcation point as an organizing centre for chaos, in: Mullin T. (ed.) The nature of chaos, Clarendon Press, London, 51-68, 1993

Murray B.T./ Chen C.F., Double-diffusive convection in a porous medium, *J. Fluid Mech.*, Vol. 201, 147-166, 1989

Muskat M./ Wyckoff R.D., An approximate theory of water-coning in oil production, *Transactions AIME*, 144-161, 1935

Nakagawa, Y., Heat transport by convection, *Physics of Fluids*, Vol. 3, No.1, 82-86, 1960

Nakanishi S./ Kawano Y./ Todaka N./ Akasaka C./ Yoshida M./ Iwai N., A reservoir simulation of the Oguni field (Japan) using MINC type fracture model, in: Barbier E./ Frye G./ Iglesias E./ Pálmason G. (eds), *Proc. of the World Geothermal Congress*, Vol. 3, 1721-1726, 1995

NAMMU - see: Atkinson e.a. 1985

Nield D.A., Surface tension and buoyancy effects in cellular convection, *J. Fluid Mech.*, Vol. 19, 341-352, 1964

Nield D.A., Onset of thermohaline convection in a porous medium, *Water Res. Res.*, Vol. 4, No.3, 553-560, 1968

Nield D.A./ Bejan A., Convection in porous media, Springer Publishers, New York, 408p, 1992

Nusselt W., Technische Thermodynamik II (Theorie der Wärmekraftmaschinen), Sammlung Göschen, Bd. 1151, 144p, 1944

Oberbeck A., Ueber die Wärmeleitung der Flüssigkeiten bei Berücksichtigung der Strömungen infolge von Temperaturdifferenzen, *Annalen der Physik und Chemie* 7, 271-292, 1879

Oertel H., Thermal Instabilities, in: Zierep/Oertel (eds), Convective Transport and Instability Phenomena, Braun Karlsruhe, 3-24, 1982

Olague N.E./ Longsine D.E./ Campbell J.E./ Leigh C.D., NEFTRAN II: User's Manual, Sandia National Laboratories, NUREG/CR-5618, Albuquerque 1991

Oldenburg C.M./ Pruess K., Dispersive transport dynamics in a strongly coupled groundwater-brine flow system, *Water Res. Res.*, Vol. 31, No.2, 289-302, 1995

Oldenburg C-M-/ Pruess K./ Travis B.J., Reply to: Comment on 'Dispersive transport dynamics in a strongly coupled groundwater-brine flow system' by Johns R.T./ Rivera A., *Water Res. Res.*, Vol. 32, No.11, 3411-3412, 1996

Oostrom M./ Hayworth J.S./ Dane J.H./ Guven O., Behaviour of dense aqueous phase leachate plumes in homogeneous porous media, *Water Res. Res.*, Vol. 28, No.10, 2123-2134, 1992

Palm E./ Tveitereid M., On heat and mass flux through dry snow, *J. Geophys. Res.*, Vol. 84, No.62, 1979

Palm E./ Weber J.E./ Kvernvold O., On steady convection in a porous medium, *J. Fluid Mech.*, Vol. 54, 153-161, 1972

Paniconi C./ Putti M., A modified Newton scheme for the solution of density dependent flow and transport equations, in: Wang S. (ed.) Proc. Adv. in Hydro-Science and - Engineering, Vol. 1, 1837-1845, 1993

Pawlowski J., Veränderliche Stoffgrößen in der Ähnlichkeitstheorie, Salle+Sauerländer, Frankfurt a.M., 108p, 1991

Pearson J.R.A., On convection cells induced by surface tension, *J. Fluid Mech.*, Vol. 4, 489-500, 1958

Peters J.H., Artificial recharge and water supply in the Netherlands, in: Johnson A.I./ Finlayson D.J. (eds), Artificial Recharge of Groundwater, ASCE, New York, 132-144, 1989

Peyret R./ Taylor T.D., Computational methods in fluid flow, Springer Publ., New York, 258p, 1985

Phillips O.M., Flow and reactions in permeable rocks, Cambridge Univ. Press, Cambridge, 285p, 1991

Piet G.J/ Zoeteman B.C.J., Bank and dune infiltration of surface water in the Netherlands, in: Asano T. (ed.), Atrificial Recharge of Groundwater, Butterworth, 529-540, 1985

Pinder G.F./ Cooper H.H., A numerical technique for calculating the transient position of a saltwater front, *Water Res. Res.*, Vol. 6, No.3, 875-882, 1970

Pinder G./ Sorek S., in: Bear J./ Cheng A. H.-D./ Herrera I./ Sorek S./ Quazar D. (eds), Seawater intrusion in coastal aquifers, Kluwer Publ., Dordrecht, to appear (1998)

Pinder G./ Stothoff S., Can the sharp interface salt-water model capture transient behaviour?, in: Celia M.A./ Ferrand L.A./ Brebbia C.A./ Gray W.G./ Pinder G.F. (eds),

Comp. Meth. in Water Res. VII, Proceedings Vol. 1, Elsevier, Amsterdam, 217-222, 1988

Platzman G.W.: The spectral dynamics of laminar convection, *J. Fluid Mech.*, Vol. 23, 481-510, 1965

Podtschaske T., Natürliche Konvektionsströmungen durch horizontale poröse Verschlüsse der Kammerzugangsstrecken in einem laugengefüllten Endlagerbergwerk, in: Hossain S./ Podtschaske T./ Rimkus D./ Stelte N./ Storck R./ Weber P. (eds), Einzeluntersuchungen zur Radionuklidfreisetzung aus einem Modellsalzstock, Projektstudien Entsorgung (PSE), Fachband 15, Berlin, 161-194, 1985

Poulikakos D., Onset of convection in a horizontal porous layer saturated with cold water, *Int. J. Heat Mass Transfer*, Vol. 28, No.10, 1899-1905, 1985

Pritchett J.W./ Garg S.K., A modeling of the Oguni geothermal field, in: Barbier E./ Frye G./ Iglesias E./ Pálmason G. (eds), *Proc. of the World Geothermal Congress*, Vol. 3, 1727-1733, 1995

Putti M./ Paniconi C., Picard and Newton linearization for the coupled model of saltwater intrusion in aquifers, *Adv. in Water Res.,* Vol. 18, No.3, 159-170, 1995

Ray D., Cellular convection with nonisotropic eddys, *Tellus*, Vol. XVII, 434-439, 1965

Rayleigh Lord, On convection currents in a horizontal layer of fluid when the higher temperature is on the under side, *Phil. Mag.*, Vol. XXXII, 529-546, 1916

Reilly T.E./ Frimpter M.H./ LeBlanc D.R./ Goodman A.S., Analysis of steady-state salt-water upconing with application at Truro well field, *Groundwater*, Vol. 25, No. 2, 194-206, 1987

Reilly T.E./ Goodman A.S., Quantiative analysis of saltwater-freshwater relationships in groundwater systems - a historical perspective, *J. of Hydrology*, Vol. 80, 125-160, 1985

Reilly T.E./ Goodman A.S., Analysis of saltwater upconing beneath a pumping well, *J. of Hydrology*, Vol. 89, 169-204, 1987

Richardson N.L., Ground water recharge at the Orange County water district, in: Johnson A.I./ Finlayson D.J. (eds), Artificial Recharge of Groundwater, ASCE, New York, 426-434, 1989

Richter W./ Flathe H., Die Versalzung von küstennahen Grundwassern dargestellt an einem Teil der deutschen Nordseeküste, in: IUGG, IAHS-Publ. No.37, 118-130, 1954

Riley D.S./ Winters K.H., Modal exchange mechanisms in Lapwood convection, *J. Fluid Mech.*, Vol. 204, 325-358, 1989

Riley D.S./ Winters K.H., Time-periodic convection in porous media: the evolution of Hopf-bifurcations with aspect ratio, *J. Fluid Mech.*, Vol. 223, 457-474, 1991

Rogers D.B./ Dreiss S.J., Saline groundwater in Mono Basin, California: 1. Distribution, *Water Res. Res.*, Vol. 31, No.12, 3131-3150, 1995a

Rogers D.B./ Dreiss S.J., Saline groundwater in Mono Basin, California: 2. Long-term control of lake salinity by groundwater, *Water Res. Res.*, Vol. 31, No.12, 3151-3169, 1995b

Rogers F.T./ Morrison H.L., Convection currents in porous media - extended theory of the critical gradient, *J. Appl. Phys.*, Vol. 21,1177-1180, 1950

Rogers F.T./ Schilberg L.E., Observation of initial flow in a fluid obeying Darcy's Law by radioactive-tracer techniques, J.Appl. Phys., Vol. 22, 233-234, 1951

Rogers F.T./ Schilberg L.E./ Morrison H.L., Convection currents in porous media-remarks on the theory, *J. Appl. Phys.*, Vol. 22, 1476-1479, 1951

Rogers F.T., Convection currents in porous media - variational form of the theory, *J. Appl. Phys.*, Vol. 24, 877-880, 1953

Roome L.R., CFEST results for HYDROCOIN level 2 case 2, Pacific Northwest Laboratories, Richland, 1989

Rubin H., On the analysis of cellular convection in porous media, *Int. J. Heat Mass Transfer*, Vol. 18, 1483-1486, 1975

Rubin H., On the application of the boundary layer approximation for the simulation of density stratified flows in aquifers, *Adv. in Water Res.*, Vol. 6, 96-105, 1983

Rubin H./ Pinder G.F., Approximate analysis of upconing, *Adv. in Water Res.*, Vol. 1, No.2, 97-101, 1977

Rubin H./ Roth C., Thermohaline convection in flowing groundwater, *Adv. Water Res.*, Vol. 6, 146-156, 1983

Ruelle D. / Takens F, On the nature of turbulence, *Comm. Math. Phys.*, Vol. 20, 167-192, 1971

Sahni B.M., Physics of brine coning beneath skimming wells, *Groundwater*, Vol. 11, No.1, 19-24, 1973

Sampson R.A., On Stokes' current function, *Phil. Trans.* A, clxxxi, 449-518, 1891

Sanford W.E./ Konikow L.F., A two-constituent solute-transport model for ground water having variable density, U.S.-Geological Survey, Water-Res. Investigations Report 85-4279, 88p, 1985

Sarker A./ Phillips O.M., Effects of horizontal gradients on thermohaline instabilities in infinite porous media, *J. Fluid Mech.*, Vol. 242, 79-98, 1992

Sauter F.J., User's manual METROPOL 1, Rijksinstitut voor Volksgezondheid en Milieuhygiene (RIVM), Interim report nr. 728514002, Bilthoven, 49p, 1987

Scheidegger A.E., General theory of dispersion in porous media, *J. Geophys. Res.* 66, 3273-3278, 1961

Schmidt R.J./ Milverton S.W., On the instability of a fluid when heated form below, *Proc. Roy. Soc.* (London) A, Vol. 152, 586-594, 1935

Schmorak S./ Mercado A., Upconing of fresh water-sea water interface below pumping wells, *Water Res. Res.*, Vol. 5, No.6, 1290-1311, 1969

Schneider K.I., Investigations of the influence of free convection on heat transfer through granular material, *11. Int. Congr. on Refrigeration*, Proc., 247-254, 1963

Schubert G./ Straus J.M., Thermal convection of water ina porous medium: effects of temperature- and pressure-dependent thermodynamic and transport properties, *J. of Geophys. Res.*, Vol. 82, No.2, 325-333, 1977

Schubert G./ Straus J.M., Transitions in time-dependent thermal convection in fluid-saturated porous media, *J. Fluid Mech.*, Vol. 121, 301-313, 1982

Schwarz H.R., Methode der finiten Elemente, Teubner Studienbücher, Stuttgart, 346p, 1984

Ségol G./ Pinder G.F./ Gray W.G., A Galerkin-finite element technique for calculating the transient position of the saltwater front, *Water Res. Res.*, Vol. 11, No.2, 343-347, 1975

Ségol G., Classic groundwater simulations - proving and improving numerical models, Prentice Hall, Englewood Cliffs, 531p, 1994

Selim H.M./ Flühler H./ Schulin R., Simultaneous ion transport and exchange in aggregated porous media, Z. dt. Geol. Ges., Vol. 136, 385-386, 1985

Senger R.K./ Fogg G.E., Stream functions and equivalent freshwater heads for modeling regional flow of variable-density groundwater 1: review of theory and verification, Water Res. Res., Vol. 26, No. 9, 2089-2096, 1990

Shen C.Y., The evolution of the double-diffusive instability: salt fingers, Phys. Fluids A1 (5), 829-844, 1989

Seipold U., The variation of thermal transport properties in the Earth's crust, J. of Geodynamics, Vol. 20, No.2, 145-154, 1995

Smith B.S., Saltwater movement in the upper Floridan aquifer beneath port royal sound (South Caolina), U.S. Geol. Survey, Water-Supply Paper 2421, 40p, 1994

Sonneveld P., CGS a fast Lancos-type solver for nonsymmetric linear systems, Report of the Dep. of Math. and Inf., No. 84-16, Techn. Univ. Delft, 18p, 1984

Sorey M.L., Numerical modeling of liquid geothermal systems, U.S.-Geol. Survey, Prof. paper 1044-D, 25p, 1978

Spink A.E.F./ Wilson E.E.M., Groundwater resource management in coastal aquifers, Int. Symp. On Groundwater Monitoring and Management, Dresden, Complex V, Paper 18, 15p, 1987

Springer J., Numerische Simulation von Porenwasserfluß und Wärmetransport mit der Methode der Finiten Elemente, Oral presentation at the Berliner Colloquium 'Modellierung in der Hydro- und Geosphäre', 1995

Springer J., private communication

Squire H.B., On the stability for three-dimensional disturbances of viscous fluid flow between parallel plates, Proc. Roy. Soc. A, Vol. 142, 621-628, 1933

Stoer J./ Bulirsch R., Einführung in die Numerische Mathematik II, Springer Publ., Berlin, 286p1973

Strack O.D.L., Some cases of interface flow towards drains, J. of Eng. Math., Vol. 6, No.2, 175-191, 1972

Strack O.D.L., A single-potential solution for regional interface problems in coastal aquifers, Water Res. Res., Vol. 12, No.6, 1165-1174, 1976

Strack O.D.L, Groundwater mechanics, Prentice Hall, Englewood Cliffs, New Jersey, 732p, 1989

Straus J.M., Large amplitude convection in porous media, J. Fluid Mech., Vol. 64, 51-63, 1974

Straus J.M.; Schubert G.: Thermal convection of water in a porous medium: effects of temperature- and pressure-dependent thermodynamic and transport properties, J. Geophys. Res., Vol. 82, 325-333, 1977

Straus J.M./ Schubert G., Three-dimensional convection in a cubic box of fluid-saturated porous material, J. Fluid Mech., Vol. 91, 155-165, 1979a

Straus J.M./ Schubert G., Three-dimensional and multicellular steady and unsteady convection in fluid-saturated porous media at high Rayleigh-numbers, J. Fluid Mech., Vol. 94, 25-38, 1979b

Straus J.M./ Schubert G., Transitions in time-dependent thermal convection in fluid-saturated porous media, *J. Fluid Mech.*, Vol. 121, 301-313, 1982

Strobl R.O./ Yeh G.T., Two-dimensional modeling of saltwater intrusion, in: Peters A./ Wittum G./ Herrling B./ Meissner U./ Brebbia C.A/ Gray W.G./ Pinder G.F. (eds), Comp. Meth. in Water Res. X, Proceedings Vol. 2, Kluwer Publ., Dordrecht, 1035-1042, 1994

Stuyfzand P.J., An accurate relatively simple calculation of the saturation index of calcite for fresh to salt water, *J. of Hydrology*, Vol. 105, 95-107, 1989

Sun Z./ Tien C./ Yen Y., Onset of convection in a porous medium containing liquid with a density maximum, *4.th Int. Conf. Heat Transfer*, Paper N.C. 2-11, Versailles, 11p, 1970

SUTRA - see: Voss 1984

Swift J.W., Hopf bifurcation with the symmetry of the square, *Nonlinearity*, Vol. 1, 333-377, 1988

SWIFT - The Intera simulator for waste injection, flow and transport, Intera Environmental Consultants 6340-016-02, Houston, 1982

Sykes J.F./ Pahwa S.B./ Ward D.S./ Lantz R.B., The validation of SWENT - a geosphere transport model, in: Stepleman R. e.a. (eds), Scientific Computing, North-Holland Publ., 351-360, 1983

Talbot C.J./ Jackson M.P.A., Salz-Tektonik, *Spektrum der Wissenschaft*, 76-86, 1987

Thiele M., Salzwasseraufstieg unter Brunnen - Ein Literaturbericht, Akademie der Wissenschaften der DDR, Inst. für Mechanik, Preprint P-Mech-08/83, 49p, 1983

Thiele M., Vermischung von Süß- und Salzwasser in horizontalen Scherströmungen im porösen Medium, Techn. Univ. Berlin, Inst. für Wasserbau und Wasserwirtschaft, Mitteilung 122, 133p, 1993

Thiele M./ Diersch H.J., 'Overshooting' effects due to hydrodispersive mixing of saltwater layers in aquifers, *Adv. Water Res.*, Vol. 9, 24-33, 1986

Thiesen/ Scheel/ Distelhorst, Wiss. Abhandlungen phys. techn. Reichsanstalt 3, 67, 1900

Tibbals C.H., Hydrology of the Floridan aquifer system in east-central Florida, U.S. Geol. Survey, Prof. Paper 1403-E, 98p, 1990

Tilton L.W./ Taylor J.K., Accurate representation of the refractivity and density of distilled water as a function of temperature, *J. of Research Nat. Bur. Stand. 18*, Paper RP971, 205-214, 1937

Tóth J., A theory of groundwater motion in small drainage basins in central Alberta, *J. of Geophys. Res.*, Vol. 67, No.11, 4375-4387, 1962

Tóth J., A theoretical analysis of groundwater flow in small drainage basins, *J. of Geophys. Res.*, Vol. 68, No.16, 4795-4812, 1963

Tóth J., The role of regional gravity flow in the chemical and thermal evolution of ground water, In: Hitchon B./ Wallick E. (eds), 1st Can./Americ. Conf. On Hydrogeology: Practical Appl. of Ground Water Geochemistry, 1984

Truesdell C.,The tragicomedy of classical thermodynamics, Springer Publ., Wien, 39p, 1971

Tsang C.F./ Buscheck T./ Doughty C., Aquifer thermal energy storage: a numerical simulation of Auburn University field experiments, *Water Res. Res.*, Vol. 17, No.3, 647-658, 1981

Turcotte D.L., Fractals and chaos in geology and geophysics, Cambridge University Press, Cambridge, 398p, 1992

Turner J.S., Buoyancy effects in fluids, Cambridge, 1973

Tyvand P.A., Thermohaline instability in anisotropic porous media, *Water Res. Res.*, Vol. 16, 325-330, 1980

de Vahl Davis G./ Mallinson G.D., False diffusion in numerical fluid mechanics, School of Mech. and Ind. Eng. Univ. of New South Wales, Report FMT/1, 22p, 1972

Valdes Gonzalez J.A., A necessary approach to the assessment of pumping groundwater resources in coastal areas, Int. Symp. on Groundwater Monitoring and Management, Dresden, Complex I, Paper 21, 26p, 1987

Veronis G., Large-amplitude Bénard convection, *J. Fluid Mech.*, Vol. 26, 49-68, 1966

Vincent A.H./ Ward D.S./ Harrover A.L., Advanced model application to the long-term waste migration from deep injection wells, Presentation at Conf. 'Solving Ground Water Problems with Models', Dallas, 1992

Vinsome P.W., ORTHOMIN an iterative method for solving sparse sets of simultaneous linear equations, 4th Symp. on Reservoir Simulation, Proc. Soc. of Petr. Eng. ,149-160, 1976

van der Vorst H., Iterative methods for the solution of large systems of equations on supercomputers, *Adv. in Water Res.*, Vol. 13, 137-146, 1990

Volker R.E/ Rushton K.R., An assessment of the importance of some parameteres for seawater intrusion in aquifers ans a comparison of dispersive and sharp-interface modelling approaches, *J. of Hydrology*, Vol. 56, 239-250, 1982

Voss C.I., SUTRA: A FE simulation model for saturated-unsaturated, Fluid-Density-Dependant Groundwater Flow with Energy Transport or Chemically-Reactive Single-Species Solute Transport, U.S. Geol. Survey, Water Resources Invest. Rep. 84-4369, 409p, 1984

Voss C.I./ Souza W.R., Variable density flow and solute transport simulation of regional aquifers containing a narrow freshwater-saltwater transition zone, *Water Res. Res.*, Vol. 23, No.10, 1851-1866, 1987

Walker K.L./ Homsy G.M., Convection in a porous cavity, *J. Fluid. Mech.*, Vol. 87, 449-474, 1978

Wang F.C., Approximate theory for skimming well formation in the Indus plain of West Pakistan, *J. of Geophys. Res.*, Vol. 70, No.20, 5055-5063, 1965

Wang M./ Bejan A., Heat transfer correlation for Benard convection in a fluid saturated porous layer, *Int. Comm. Heat Mass Transfer*, 14, 617-626, 1987

Ward D.S./ Buss R.B./ Morganwalp D.W./ Wadsworth T.D., Waste confinement performance of deep injection wells, Presentation at Conf. 'Solving Ground Water Problems with Models', Denver, 1987

Weber J.E., The boundary-layer regime for convection in a vertical porous layer, *Int. J. Heat Mass Transfer*, Vol. 18, 569-573, 1975

Welsh J.L., Ground water management: Los Angeles coastal plan, in: Johnson A.I./ Finlayson D.J. (eds), Artificial Recharge of Groundwater, ASCE, New York, 416-425, 1989

Welty C./ Gelhar L.W., Stochastic analysis of the effects of fluid density and viscosity variability on macrodispersion in heterogeneous porous media, *Water Res. Res.*, Vol. 27, 2061-2075, 1991

Welty C./ Gelhar L.W., Simulation of large-scale transport of a variable density and viscosity fluids using a stochastical mean model, *Water Res. Res.*, Vol. 28, 815-827, 1992

Wilson J.L./ Sa da Costa A., A numerical model of seawater intrusion in aquifers, MIT Sea Grant College Program Report No. 79-27, 1979

Wilson J.L./ Sa da Costa A., Finite element simulation of a saltwater/freshwater interface with indirect toe tracking, *Water Res. Res.*, Vol. 18, No.4, 1069-1080, 1982

Wirojanagud P./ Charbeneau R.J., Saltwater upconing in unconfined aquifers, *J. of Hydr. Eng.*, Vol. 111, No.3, 417-434, 1985

Wolff J., Umweltgerechte Grundwasserbewirtschaftung in ökologisch sensiblen Bereichen der Nordseeküste, Oral presentation at 'Kolloquium - Inst. für Angewandte Geowissenschaften', Techn. Univ. Berlin, 1997

Wooding R.A., Steady state free thermal convection of liquid in a saturated permeable medium, *J. Fluid Mech.*, Vol. 2, 273-285, 1957

Wooding R.A., The stability of a viscous liquid in a vertical tube containing porous material, *Proc. Roy. Soc. A*, Vol. 252, 120-134, 1959

Wooding R.A., Free convection of fluid in a vertical tube filled with porous material, *J. Fluid Mech.*, Vol. 13, 129-144, 1962

Wooding R.A., Convection in a saturated porous medium at large Rayleigh number or P'eclet number, *J. Fluid Mech.*, Vol. 15, 527-544, 1963

Yano Y., A practical procedure for vertical two-dimensional hydrothermal simulation by finite element method, *J. Jap. Assoc. Petrol. Eng.*, Vol. 54, 18-31, 1989

Yen Y., Effects of density inversion on free convective heat transfer in porous layer heated from below, *Int. J. Heat Mass Transfer* 17, 1349-1356, 1974

Yusa Y., Numerical experiment of groundwater motion under geothermal condition - vying between potential flow and thermal convective flow, *J. of the Geothermal Res. Soc. of Japan*, Vol. 5, No.1, 23-38, 1983 (in Japanese)

Yusa Y./ Oishi I., Theoretical study of two-phase flow through porous medium (II), *Journal of the Geothermal Research Society of Japan*, Vol. 11, No.3, 217-237, 1989

Zebib A./ Kassoy D.R.: Onset of natural convection in a box of water-saturated porous media with large temperature variation, *Physics of Fluids*, Vol. 20, 4-9, 1977

Zebib, A./ Kassoy, D.R. Three-dimensional natural convection motion in a confined porous medium, *Physics of Fluids*, Vol. 21, 1-3, 1978

Zheng C./ Bennett G.D., Applied contaminant transport modeling, Van Norstrand Reinhold, New York, 440p, 1995

Ziagos J.P./ Blackwell D.D., A model for the transient temperature effects of horizontal fluid flow in geothermal systems, *J. of Volcanology and Geothermal Res.*, Vol. 27, 371-397, 1986

Appendix I: FAST-C(2D) Input- and Output-Files

Input-File for FAST-C(2D)

```
Fast Algorithm for Saline and Thermal
            Convection
          in porous media
developed by E. Holzbecher version 2/98
```

INPUT FILE SPECIFICATIONS

```
-------------------INPUT CARDs 0-------------
Version Identifier (CHARACTER*1,2 INTEGER)
                            FORMAT (A1,I1,I2)
Title (CHARACTER)           FORMAT (*)
```

First line is 'X1 1' for input files of current version.
Maximum length of 'title' is 100 characters.

```
------------------INPUT CARD 1--------------
NA,NX,IEXPA,NZ (4 INTEGER)
```

NA   Max. no. of characteristics in inhomogeneous
                             array input - default: 16
NX   No. of blocks in x-direction
NZ   No. of blocks in z-direction(direction of gravity)
IEXPA   Nonlinear thermal expansion characteristics
            >0   constant expansion coefficient
            <=0   variable expansion coefficient
            =0   min. temperature above 4°C
            =-1   min. temp. below 4°C,
                                    max. temperature above 4°C
            =-2   max. temperature below 4°C

(calculated from this values is N=NX*NZ= total no. of blocks)

```
------------------INPUT CARD 2--------------
(DX(I),I=1,NX)                    FORMAT (*)
 DY(1)
(DZ(I),I=1,NZ)        (all REAL)
```

Blocklengths in both coordinate directions, (DY not used)

```
------------------INPUT CARD 3--------------
ITYP,IRST,JRST,LOUT,ISTD,ITME,IUNIT,IBOUS
(8 INTEGER)
```

              ITYP   Choice of convection type
                              (0: saline, 1: thermal)
              IRST   Set equal 1 for restart job
              JRST   Set equal 1 to enable restart in.
                            additional job
              LOUT   Output control parameter
                              (0: maximum, 9: minimum output)
              ISTD   Steady state option
                              (0: transient, 1: steady state)

ITME   No. of time periods to be simulated (<7)
       (see CARD 13)
IUNIT   Option 'dimensionless' / ' physic. units'
       (=1: dimensionless, =0: phys. units)
IBOUS   not used in current version

`------------------INPUT CARD 4--------------`
`DRHO,TMIN,TMAX,GRAV,RAL (5 REAL)`

DRHO   Fluid density change per unit
      (saline: increase, thermal: decrease)
      (not used if IEXPA<=0 or RAL>0)
TMIN   Minimum temperature in K
TMAX   Maximum temperature in K
      (not used if: IEXPA<=0 and RAL>0)
GRAV   Gravity (not used if RAL>0)
RAL   Rayleigh-number
      (only used if IUNIT=1, see CARD 3)

`------------------INPUT  CARD  5--------------`
`VISC, DVISC, IVISC (2 REAL, 1 INTEGER)`

VISC   Reference dynamic fluid viscosity
      (not used if RAL>0)
DVISC   Viscosity change
      (set zero for constant viscosity case)
IVISC   Viscosity change parameter
      (=0: linear , =1: nonlinear change)

`------------------INPUT CARD 6--------------`
`ISPACE,ICORR,IGEOM,ALPHA,ORIGIN`
`                         (3 INTEGER, 2 REAL)`

ISPACE =2 BIS discretization (upwind scheme)
       =1 CIS discretization (central in space)
ICORR   correction of numerical dispersion

IGEOM   Geometry choice (=0: cartesian, =1: radial)
ALPHA   Opening angle (only for radial geometry)
ORIGIN Origin location (only for radial geometry)

```
------------------INPUT CARD 7--------------
Array input for variables:
```

Initial streamfunction   (only if not RESTART)
Initial concentration
            (only for saline cases and if not RESTART)
Initial temperature
            (only for thermal cases and if not RESTART)
X-direction permeability (used only for IUNIT=0, see CARD 4)
Diffusivity
Z-direction permeability (used only for IUNIT=0, see CARD 4)
Heat capacity ratio (used only in thermal cases)
Retardation (used only for transient saline cases)
Porosity (used only for transient saline cases)
Salt source/sinkrate (for saline cases)
Longitudinal dispersivity
Transversal dispersivity
Heat source/sinkrate (for thermal cases)

for transient cases:
```
(TS(I),I=NA),(TT(I),I=1,NA)  (2 INTEGER ARRAYs)
```
            time characteristic for salt source/sinkrates
for steady cases:
```
TS(1),TT(1)  (2 INTEGER)
```
            not used in current version

Characteristic for a source or sink in a transient model is determined
by adding the characteristic numbers for the timeperiods with non-
zero re-/discharge. The characteristic numbers for timeperiods are:

Timeperiod	1	2	3	4	5	6
Character. number	1	2	4	8	16	32

Table: Characteristic numbers for timeperiods

Example: if a well is pumped in timeperiods 1,2 and 4, then input: 11.

array (X) input is:
> I1   (1 INTEGER) array input control parameter

if I1<0: (read from file TAPE4)
> READ (4,*) (X(I),I=1,N) (REAL ARRAY)

if I1=0: (inhomogeneous case)
> (W(I),I=1,NA)
>> variable values used in the model (
>> (for all parameters except source/sink-rates)

or: ((W(I,J),I=1,NA),J=1,6) (REAL ARRAY)
>> (for source/sink-rates)
> (K(I),I=1,N) (INTEGER ARRAY)
>> location of variable values in model

if I1>0: (homogeneous case)
> X (1 REAL) constant within entire model area

------------------INPUT CARD 8---------------
Streamfunction boundary conditions

Boundary condition for a variable is:
> boundary condition for 4 edges:
>> 1. constant x-edge (x-block length DX(1))
>> 2. constant x-edge (x-block length DX(NX))
>>> constant y-edges (not used)
>> 3. constant z-edge (z-block length DZ(1))
>> 4. constant z-edge (z-block length DZ(NZ))

Boundary condition for one edge is:
> IB (1 INTEGER) boundary condition homogeneity switch

if IB>0   (homogeneous case)
> IBB,S (1 INTEGER, 1 REAL)
>> Boundary condition characteristics for whole edge

if IB0=0   (inhomogeneous case)
                (IBB(I),I=1,NW) (INTEGER ARRAY)
         (S(I),I=1,NW) (REAL ARRAY) (NW is NX or NZ)
                     boundary condition characteristic for single block

Boundary condition characteristics are:
         IBB   Boundary condition type parameter
                          =3 NEUMANN boundary condition
                          =2 DIRICHLET boundary condition
                          =1 DIRICHLET boundary condition with S=0.
         S   Value of boundary condition

------------------INPUT CARD 9---------------
Transport boundary conditions

boundary conditions as for streamfunction, here for temperature (in thermal case) or concentration (in saline case)

         IBB   =3   DIRICHLET boundary condition
                =2   DIRICHLET boundary condition for input only
                =1   NEUMANN boundary condition (=0)

------------------INPUT CARD 10a------------
ICG,CN2,OMS,OMT,EPS1,EPS2,EPS3,MAX1,MAX2,MAX3
                  (1 INTEGER, 6 REAL, 3 INTEGER)

     ICG   Choice of solver for transport equation
                      (=0: conjugate gradient CG, =1: CGS,
                                  =2: Bi-CG, =3: BiCGstab)
     CN2   Timelevel weighting factor   =0. totally implicit,
                                      =.5 CRANK-NICOLSON,
                                      =1. explicit)
     OMS   Preconditioning parameter for flow equation
                      (=0.0: no preconditioning; <0.0 scaling)
     OMT   preconditioning parameter for transport equation

(=0.0: no preconditioning; <0.0 scaling)

EPS1	Accuracy for solvers of linear systems
EPS2	Accuracy heat/salt to end Picard iteration
EPS3	Accuracy streamfunction to end Picard it.
MAX1	Maximum of iterations in Picard iteration
MAX2	Maximum of iterations to solve transport
MAX3	Maximum of iterations to solve flow

```
------------------INPUT CARD 10b------------
IDBL,INORM,ICRT                    (3 INTEGER)
```

IDBL	Switch double/single option
	(=1: double precision/ =0: single precision)
INORM	Switch norm option
	(=2: 2-norm, else: maximum-norm)
ICRT	Switch criterion type option
	(=0: residual type, =1: change type)

```
------------------INPUT CARD 11-------------
(TIMS(I),I=0,6),(NDT(I),I=1,6),(BETA(I),I=1,6)
                    (7 REAL, 6 INTEGER, 6 REAL)
```

(only in transient simulation)

TIMS(0)	initial time
TIMS(I)	end of time period I
	(= begin of time period I+1)
NDT(I)	>0: no. of timesteps in timeperiod I
	<0: automatic timestepping
	=-1 to fulfill NEUMANN criterion
	=-2 to fulfill COURANT criterion
	=-3 to fulfill NEUMANN & COURANT criterion
BETA(I)	automatic timestepping parameter for timeperiod I

```
-------------------INPUT CARD 12--------------
IPRS,IPRC,IPRV,IPRN,IPRM,INEC,IPRI,IPRP,IPCO,
IF1,IF2,IF3
JPRS,JPRC,JPRV,JPRN,JPRM,JNEC,JPRI,JPRP,JPCO,
JF1,JF2,JF3
                              (2 X 12 INTEGER)
```

IPRS	output streamfunction to TAPE7 (TAPE8)
IPRC	" temperature/concentration (TAPE9)
IPRV	" velocities (TAP10)
IPRN	" NUSSELT-numbers (TAPE11)
IPRM	" maximum values velocity and streamfunction (TAPE12)
INEC	" eddy characteristics (TAPE13)
IPRI	" transport var. min./max./medium (TAPE14)
IPRP	" grid PECLET-numbers (TAPE15)
IPCO	" COURANT-numbers (TAPE16)
IF1	" fluid mass balance (TAPE17)
IF2	" transport variable flux balance (TAPE*)
IF3	" viscosity

J***   same as I*** for special file output (TAPE8)

Output is done:
   - after end of calculation in steady state case, if parameter>0
   - in every time-period after each I*** (resp. J***) iterations
     in transient simulation

------------------INPUT CARD 13-------------
```
IPL,IPL1,IPL2,IPL3,IPL4,IPL5,IPL6,IPL7,PLEN,
PWID
```
(8 INTEGER, 2 REAL)

IPL	=1 initiates plot output (for steady state; for transient: after each IPL iterations)
IPL1	no. of isolines for temperature resp. concentr.
IPL2	no. of isolines streamfunction
IPL3	graphics mode (see: Table)
IPL4	hardcopy question
IPL5	linetype
IPL6	plot grid
IPL7	automatic scaling for isoline levels
PLEN	length of picture (max. 1.6)
PWID	width of picture (max. 1.0)

------------------INPUT CARD 14-------------
```
KPRT1,KPRT2,KPRT3,KPRT4,KPRT5,KPRT6
```
(6 INTEGER)

File output format options:

KPRT1	=0: hexadecimal; else: ASCII
KPRT2	>0: output number of blocks
	<0: output SURFER-readable
KPRT3	>0: output blocklengths
KPRT4	>0: output x-coordinate
KPRT5	>0: not used in FAST-C(2D)
KPRT6	>0: output z-coordinate

Table: FAST graphics modes

FAST Graphic modus	Display size (Pixel)	Colours	Graphic type
1	320 x 200	16 colours	EGA
2	640 x 200	16 colours	EGA
3	640 x 350	4 or 16	EGA
4	640 x 480	black & white	VGA
5	640 x 480	16 colours	VGA
6	320 x 200	256 colours	VGA
7	640 x 480	256 colours	SVGA
8	800 x 600	16 colours	SVGA
9	800 x 600	256 colours	SVGA
10	1024 x 768	16 colours	SVGA
11	1024 x 768	256 colours	SVGA
12	1280 x 1024	16 colours	SVGA
13	1280 x 1024	256 colours	SVGA
14	all		
15	all		

Table: Graphics modes used by FAST

Output-Files

TAPE7 - Standard- Outputfile

TAPE8 - Outputfile for calculated streamfunction arrays

```
X, Z, S                    (3 FLOAT in FORMAT (3E15.8))
```

X- resp. Z -coordinates and value of streamfunction

(Arrays for different timesteps are written after each other in the file)

TAPE9 - Outputfile for calculated arrays of the transport variable

```
X, Z, T                    (3 FLOAT in FORMAT (3E15.8))
```

X- resp. Z -coordinates and value of transport variable temperature or salinity

(Arrays for different timesteps are written after each other in the file)

TAPE10 - Outputfile for calculated arrays of velocity-components

```
(VX(I),I=1,N),  (VZ(I),I=1,N)
                      (FLOAT-arrays in FORMAT (8E15.8))
```

arrays containing X- resp. Z -components of velocity vectors

(Arrays for different timesteps are written after each other in the file)

TAPE11 - Outputfile for Nusselt- resp. Sherwood-numbers

```
T, QU1,QD1,QU2,QD2,QU3,QD3,QL1,QR1,QL2,QR2,QL3,QR3
                      FORMAT (8E15.8, 12(1X,E10.5)
```

with T   simulation time
      QU1   1.order approx. of Nusselt- resp. Sherwood-no. for top boundary
      QD1   1.order approx. of Nusselt- resp. Sherwood-no. for bottom
      QU2   2.order approx. of Nusselt- resp. Sherwood-no. for top boundary

QD2   2.order approx. of Nusselt- resp. Sherwood-no. for bottom
QU3   3.order approx. of Nusselt- resp. Sherwood-no. for top boundary
QD3   3.order approx. of Nusselt- resp. Sherwood-no. for bottom
QL1   1.order approx. of Nusselt- resp. Sherwood-no. for left boundary
QR1   1.order approx. of Nusselt- resp. Sherwood-no. for right bound.
QL2   2.order approx. of Nusselt- resp. Sherwood-no. for left boundary
QR2   2.order approx. of Nusselt- resp. Sherwood-no. for right bound.
QL3   3.order approx. of Nusselt- resp. Sherwood-no. for left boundary
QR3   3.order approx. of Nusselt- resp. Sherwood-no. for right bound.

TAPE12 - Outputfile for maxima of velocity and streamfunction extrema

```
T, VM, I1, J1, SMIN, I2, J2, SMAX, I3, J3
                    FORMAT (E15.8,3(E15.8,2I4))
```

with T   simulation time
      VM   maximum of absolute value of velocity
      I1, J1 block-indices corresponding with VM
      SMIN   minimum of absolute value of streamfunction
      I2, J2 block-indices corresponding with SMIN
      SMAX   minimum of absolute value of streamfunction
      I3, J3 block-indices corresponding with SMAX

TAPE13 - Outputfile for eddy characteristics

```
T, S, I, J (for first entry) resp. S, I, J (for following entries)
  FORMAT (2E12.5,I4,I4) resp. FORMAT (12X,E12.5,I4,I4)
```

with T   simulation time
      S   value of local extremum of streamfunction
      I, J block indices corresponding with local extremum S

TAPE14 - Outputfile for extrema and medium values of transport variable

```
T, CMAX, CMIN, C1, C2, C3, C4    FORMAT (7E15.8)
```

with T   simulation time
      CMAX maximum of. transport variable
      CMIN minimum of transport variable.
      C1   medium of transport variable near left boundary
      C2   medium of transport variable near right boundary
      C3   medium of transport variable near upper boundary
      C4   medium of transport variable near bottom

TAPE15 - Outputfile for maxima of grid Peclet-numbers

```
T, PEX, PEZ                        FORMAT (3E15.8)
```

with T   simulation time
    PEX   maximum of x-direction grid-Peclet-no.
    PEZ   maximum of z-direction grid-Peclet-no.

TAPE16 - Outputfile for maxima of Courant-numbers

```
T, COX, COZ                        FORMAT (3E15.8)
```

with T   simulation time
    COX   maximum of x-direction Courant-no.
    COZ   maximum of z-direction Courant-no.

TAPE17 - Outputfile for fluid mass balance

```
T, QF1, QF2, QF3, QF4              FORMAT (5E15.8)
```

with T   simulation time
    QF1   balance of flux terms across left boundary
    QF2   balance of flux terms across right boundary
    QF3   balance of flux terms across top boundary
    QF4   balance of flux terms across bottom boundary

TAPE18 - Outputfile for flux balance of the transport variable

```
T, QT1, QT2, QT3, QT4              FORMAT (5E15.8)
```

with T   simulation time
    QT1   balance of flux terms of transport variable across left boundary
    QT2   balance of flux terms of transport variable across right boundary
    QT3   balance of flux terms of transport variable across top boundary
    QT4   balance of flux terms of transport variable across bottom

TAPE19 - Outputfile for calculated arrays of viscosity

```
X, Z, S                     (3 FLOAT in FORMAT (3E15.8))
```

X- resp. Z -coordinates and value of viscosity

(Arrays for different timesteps are written after each other in the file)

Index

A

accuracy criterion 66, 105, 142

advection 7, 27, 31, 33, 49, 50, 68, 216, 228

anisotropy 8, 84, 85, 97, 99, 126, 225

B

Bénard convection 3, 8, 9, 33, 75-78, 91, 101, 161, 176

benchmark 6, 8, 99, 114, 173

bifurcation 88, 89, 94, 126, 129, 130, 131, 176
  Hopf 129-131, 176
  pitchfork 88, 89, 129, 130

boundary conditions 6, 9, 35, 49, 58-61, 78-80, 126, 150, 162, 165, 173, 176, 180, 181, 196, 199, 202, 217, 231, 232, 236-238, 244-246, 273, 274
  Cauchy 61
  Dirichlet 50, 60, 61, 149, 162, 173, 196, 217, 218, 231, 233, 245, 246

buoyancy 2, 32, 33, 37, 41, 77, 126, 208, 220

C

calibration 6, 35

chaos 89, 130

chord slope method 64

conduction 27, 30, 31, 46, 80, 82, 88, 92, 106, 115, 123, 129, 130, 150-153

conductivity XVI, 11, 17-20, 25, 31, 35, 38, 46, 125, 161, 182, 200, 209-211, 221, 222, 228
  hydraulic 35, 38

thermal XVI, 11, 18, 19, 25, 31, 46, 228

conjugate gradients 68-71, 98, 138, 141-144, 155, 236, 274

conservation 21, 26-28, 30, 31, 39, 118
  energy 30, 118

continuity equation 32, 37, 38, 44, 195

convection IX, XVI, 3, 8, 9, 27, 31, 33, 43, 75-81, 84-92, 95, 96, 98, 101, 104, 107, 109, 111-118, 121-123, 125-127, 129-131, 133, 134, 141, 149-151, 161, 168, 170, 171, 173-175, 182, 185-188, 195, 198, 233, 240, 241, 244, 270

convergence
  of numerical algorithms 62, 63, 68, 71, 72, 97

coordinates
  cylindrical 152, 153

Courant-criterion XVII, 58, 59, 90, 97, 98, 105, 111, 133, 135, 140, 163, 187, 188, 224, 236, 281

D

Darcy's law XVII, 1, 26, 29, 32-35, 37-39, 44, 125, 161, 195

density IX, XVI, XVII, 1-4, 7-9, 11, 13-18, 20-23, 25-28, 31, 32-34, 36, 37, 39-44, 49-51, 61, 73, 76, 78, 99, 109, 112, 118-122, 126, 149, 150, 171-173, 179, 181, 182, 191, 193, 195, 201, 207, 208, 216, 217, 219-222, 228-232, 238, 243, 247, 248, 251, 271
  fluid XVI, 1, 2, 13, 21, 23, 31, 120, 171, 193, 220, 222, 231

E

F

G

H

I

L

M

N

Y

Printing: Mercedesdruck, Berlin
Binding: Buchbinderei Lüderitz & Bauer, Berlin